HUMAN PHYSIOLOGY, BIOCHEMISTRY AND BASIC MEDICINE

HUMAN PHYSIOLOGY, BIOCHEMISTRY AND BASIC MEDICINE

LAURENCE A. COLE
PETER R. KRAMER

ELSEVIER

AMSTERDAM • BOSTON • HEIDELBERG • LONDON
NEW YORK • OXFORD • PARIS • SAN DIEGO
SAN FRANCISCO • SINGAPORE • SYDNEY • TOKYO
Academic Press is an imprint of Elsevier

AP

Academic Press is an imprint of Elsevier
125 London Wall, London, EC2Y 5AS, UK
525 B Street, Suite 1800, San Diego, CA 92101-4495, USA
225 Wyman Street, Waltham, MA 02451, USA
The Boulevard, Langford Lane, Kidlington, Oxford OX5 1GB, UK

Notices

Knowledge and best practice in this field are constantly changing. As new research and experience broaden our understanding, changes in research methods, professional practices, or medical treatment may become necessary.

Practitioners and researchers must always rely on their own experience and knowledge in evaluating and using any information, methods, compounds, or experiments described herein. In using such information or methods they should be mindful of their own safety and the safety of others, including parties for whom they have a professional responsibility.

To the fullest extent of the law, neither the Publisher nor the authors, contributors, or editors, assume any liability for any injury and/or damage to persons or property as a matter of products liability, negligence or otherwise, or from any use or operation of any methods, products, instructions, or ideas contained in the material herein.

Library of Congress Cataloging-in-Publication Data
A catalog record for this book is available from the Library of Congress

British Library Cataloguing in Publication Data
A catalogue record for this book is available from the British Library

For information on all Academic Press publications
visit our website at http://store.elsevier.com/

ISBN: 978-0-12-803699-0

Publisher: Mica Haley
Acquisition Editor: Stacy Masucci
Editorial Project Manager: Sam Young
Project Manager: Edward Taylor
Designer: Matthew Limbert

Contents

3
HUMAN ORGANS

3.1 The Endocrine System

3.2 The Heart and Lungs

3.3 The Circulation

3.4 Brain and Nervous System

3.5 Seeing, Hearing, Tasting, Smelling, and Touching

3.6 The Liver and Digestive System

3.7 The Kidney Filtration System

3.8 Sex and Reproductive Systems: Male

3.9 Sex and Reproductive Systems: Female

Preface

How this book was conceived and came about is in many respects a book unto itself. Laurence Cole, PhD, an author, was a professor of obstetrics and gynecology at University of New Mexico and an affiliate and associate professor at Yale University. He was a scientist and a medical student reproductive biology teacher. He was chief of the division of women's health research.

Dr. Cole started the USA hCG Reference Service, a small business run from his new home in Angel Fire, in the mountains of New Mexico. The USA hCG Reference Service ran blood and urine laboratory tests and consulted with physicians worldwide in cases of gestational trophoblastic disease and cases of women not-pregnant positive in hCG gestational tests. The USA hCG Reference Service only occupied Laurence's time for just 1 day in each 14 days.

Laurence could not stand all the free time he had, since, in essence for 13 of 14 days, he was doing nothing, other than just standing and staring at the incredible mountains surrounding Angel Fire. Laurence was so used to thinking about reproductive science and dreaming reproductive science that he sat by the computer and started writing. In 2012 and 2013, Laurence wrote a comprehensive physiology and medicine textbook on human reproduction for medical students entitled "The Biology and Medical Dynamics of Human Reproduction" Nova Science Publishers 2014, ISBN 978-1-62948-832-5. In 2013, Laurence wrote other medical books, including "The Glycopeptins," Nova Science Publishers, 2014 ISBN 978-1-63321-833-8.

In 2014, Laurence kept writing, preparing the second edition of the monograph "Human Chorionic Gonadotropin (hCG)," Elsevier Insight, ISBN 978-0-12-384907-6. Next, Laurence prepared talks on human biology for presentation at Texas high schools, which is when he considered writing a high school human biology textbook. He proposed this book to Nova Science Publishers. They responded with a solid No!, saying they do not publish high school textbooks. Laurence then proposed the same high school science book to Elsevier. Elsevier also said No!, saying they only publish university undergraduate and graduate school books, but instead suggested an undergraduate and graduate student book on human biology.

Angel Fire is located at 8400 ft altitude. In the summer, the temperature rarely exceeds 75 degrees. However, the sun is different at 8400 ft altitude. The solar rays or infra-red heat is very much more intense in Angel Fire, such that 75 degrees feels on the skin like 90-100 degrees. I had sat outside in the sun at our house for 1 h and my forehead felt like it was on fire. I then went inside to get away from the sun and sat alone on the sofa. That is when I suddenly came up with the book title humanology, a book that combined all aspects of physiology with biochemistry and anatomy, or a book that looks at every conceivable aspect of what makes a human. I was sold. I wrote to Elsevier proposing the book "Humanology," a comprehensive book covering all -ologies of humans. A new book for undergraduate students at university.

What is humanology? What does the word mean? I was able to find three dictionaries that listed the word humanology. What was found, however, what a mishmash of definitions with different biological and psychological meanings.

The Merriam Webster Dictionary, http://nws.merriam-webster.com.

Humanology (noun): the balance between the wealth mindset and the spiritual self. The science that deals with human aspects for betterment of relationships and self-confidence. The process and use of the mental, emotional, physical and spiritual states of a human being.

The Syntheory Dictionary, http://www.syntheory.com/refs/glossary.html.

Humanology—the current author defines humanology as the scientific study of humanity. It draws upon all of the existing theories of psychology as well as all of the known arts and spiritualities.

The Wikipedia Encyclopedia, http://philosophyofreason.wikispaces.com.

Humanology is the study of human beings. It is a neologism used within POR to identify the difficulty of using science to study human beings. Humanology consists of all serious attempts to apply scientific reasoning to Homo sapiens. By definition, humanology includes psychology, sociology, anthropology, and all similar academic specialties.

I start this book with one simple definition on humanology: Humanology is one science describing the actions, all physiological and biochemical behaviors, and an introduction to medical behavior of humans. It is the science of everything human.

I soon started to realize that I needed partners for writing this book, my co-authors or partner authors. Partners with expertise in biochemistry, human physiology, and human psychology. I thought about Peter Kramer, PhD, a nutritional biochemist. Laurence Cole and Peter Kramer both studied together on their biochemistry PhDs at Medical College of Wisconsin in Wauwatosa, Wisconsin, 1978-1982. I always remember that Christmas when Peter and I, who both drove very old Volkswagen Beetles at that time, said that we will meet for lunch down at the tavern on Wisconsin Avenue. Wisconsin Avenue in Wauwatosa is very hill-ridden with dozens of ups and downs. There was at least a foot of fresh snow on the ground and no traffic. Stephen left the medical complex and went over the first hill on a car-deserted Wisconsin Ave and unwittingly his car rotated around six or seven times coming over the brim of the first hill, skiing in every direction.

I followed him, lost all control, and did the same thing coming over the hill spinning around and around. My spinning car slapped side into side into Peter's car as I came down the hill. I remember the front door of Peter's car falling off, tumbling flat into the snow, then a few minutes later the back door falling off. Then 10 min later, his front wheel dropped 90 degrees, and finally his back wheel falled flat 90 degrees. All I could do is laugh. All we all could do is laugh and laugh as we walked into the tavern for lunch. Peter's old beetle had had it.

I remember that after lunch and drinking a couple of beers, coming outside and looking at the cars, all we could do is laugh more looking at Peter's door-less and wheel-less dead skeleton car. We ordered taxis to take us home. This story here is one of unraveling a car, yet here we write a book on unraveling humans piece by piece.

Stephen seemingly traced Laurence all across the United States. Laurence Cole first moved east to University of Michigan in Ann Arbor, Michigan. Stephen moved east to Notre Dame University in Notre Dame, Indiana. Laurence moved further east to Yale University in Connecticut, and Stephen moved east shortly afterward to University of Connecticut. We always kept in touch, worked together, and hiked together on weekends. Now Stephen lives in the Philippines where he co-authors this book by e-mail with me.

This comprehensive text is designed to give the undergraduate student a thorough understanding of the workings of a human being, encompassing the workings of the brain and mind, the workings of the organs or body, the underlying biochemistry, and a broad introduction to medicine. Such knowledge is essential to a career in the medical field.

The book starts by looking at the bodily units (Section 1: Human Body Formed by Units). Here we examine the underlying cells and inspect the biochemistry at the roots of the functioning of the human body. We also investigate the underlying DNA and genetics of the human body and how it all works.

The book then examines the anatomy and biology of the human frame (Section 2: Human Frame). First, by examining the human skeleton, then inspecting the skin, ligaments, and muscles of the human body. Third, this section examines apoptosis and the underlying mechanism by which we grow and age. The third section is a comprehensive examination of human physiology (Section 3: Human Organs), looking at system by system how the body functions and what different organs look after.

The fourth section of the book examines in detail how humans evolved or came about (Section 4: Human Evolution and Pseudoevolution). It looks at how humans evolved from primates, how Homo sapiens first appeared, including details about early life and how it slowly became civilized. Section 5 deals with human nutrition, diet, and fitness and obesity (Section 5: Nutrition).

Section 6 is an introduction to basic human medicine (Section 6: Human Diseases and Treatments). This section selects multiple categories of disease, genetic defects, infection, cancer, cardiovascular disease, hydatidiform mole, diabetes, and ovarian cysts. It tries to introduce each disease category, examine its underlying physiology and biochemistry, and explain the basic mechanisms of disease.

It is believed that a comprehensive understanding of all six sections of this book will provide students with complete knowledge of the workings of human beings, or an appropriate introduction with the vast science of humanology.

Laurence A. Cole, PhD
Peter R. Kramer, PhD

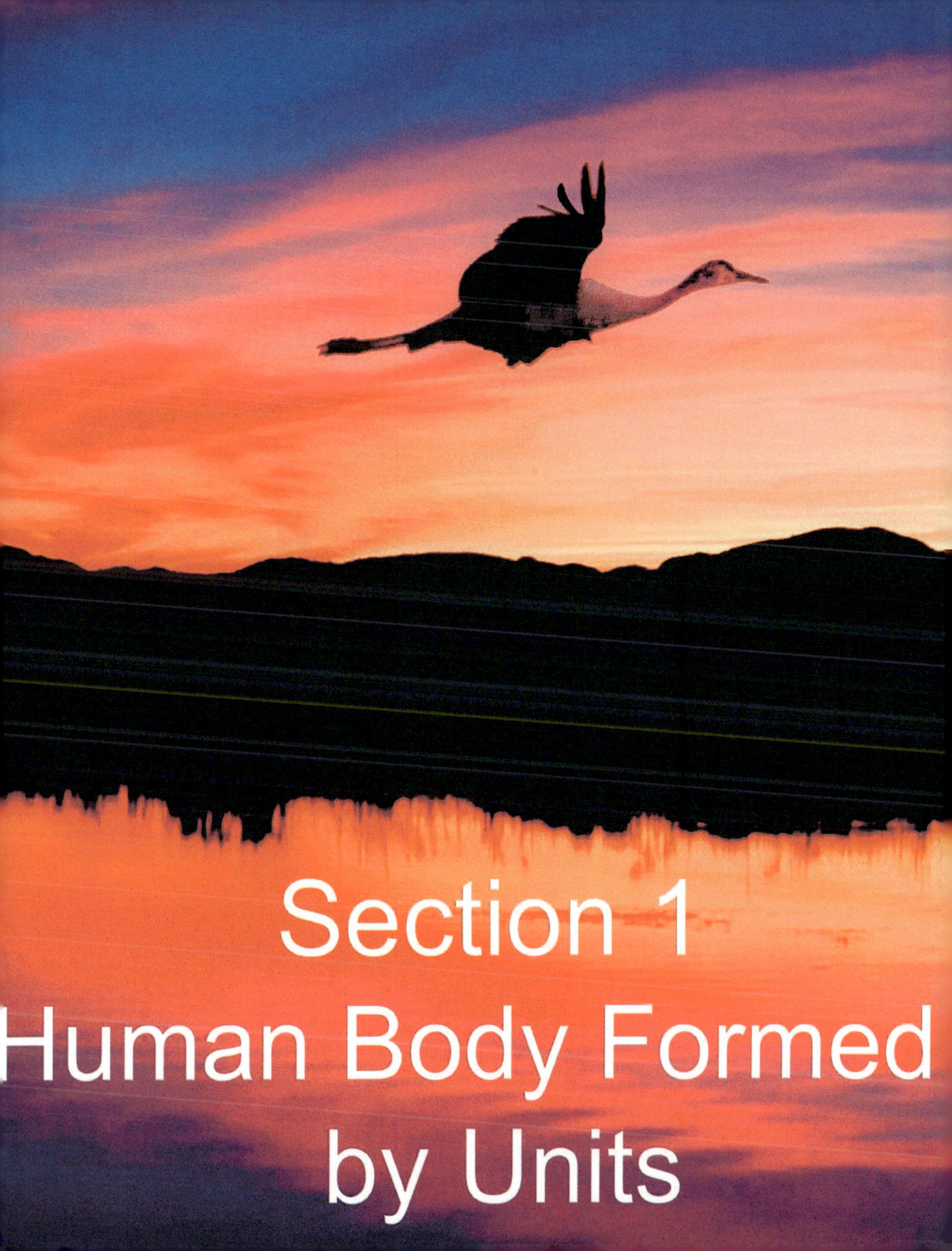

Section 1
Human Body Formed by Units

CHAPTER

1.1

The Cell

The cell is the unit of all animal and plant, prokaryotic and eukaryotic, life. There is single cell life like the amoeba, oligocell life like some plankton, and multicell life like humans. Humans actually contain 3.0×10^{13} cells or 30 trillion cells. If you imagine that an average speck of dust weighs just 0.75 μg, then the average 70 Kg man comprises the weight of 93 billion particles of dust. Considering both values, the average cell in the human body weighs just a tiny fraction of the mass of a speck of dust, 1/323 of a speck of dust. The cell is the basic human makeup or unit, a tiny fraction of a speck of dust.

This miniscule body, a fraction of the size of a speck of dust, is an entire village full of organelles or shops and business facilities. There is the mitochondrion, which is like the village power station, the nucleus, where the village mayor controls the village council or chromosomes and orders what is to be made and when, the rough endoplasmic reticulum and ribosomes, which are like the local village factory, hard at work manufacturing the village's need for proteins, and the Golgi apparatus, another village factory, hard at work refining the synthesized proteins. Then there is the lysosomes or local village dump degrading the molecules not needed by the cells. All buildings in the village or fraction of a speck of dust work happily and eagerly together.

Figure 1.1.1 shows a representative human cell. While there are thousands of types or classes of cells in humans, each is different in shape, makeup, and function. A very general cell is shown in Figure 1.1.1. The number of each organelle varies in each cell type, and some cell types contain specialized functional organelles, which are not shown. Each cell is centered by a nucleus and within each cell is a nucleolus. Each cell has a smooth and rough endoplasmic reticulum, ribosomes, Golgi apparatus, mitochondria, lysosomes, centrioles, and microtubules. Shown are secretory vesicles emerging from the Golgi apparatus as a representative cell (Figure 1.1.1).

THE PLASMA MEMBRANE

The plasma membrane is the envelope that encloses the entire cell. The envelope is made up of molecules called phospholipids. These are fatty acids with a charged or polar group on the head, and two long noncharged fatty tails comprising carbon and hydrogen atoms. Many different lengths of fatty acid tails, saturated

Human Physiology, Biochemistry and Basic Medicine
http://dx.doi.org/10.1016/B978-0-12-803699-0.00001-3

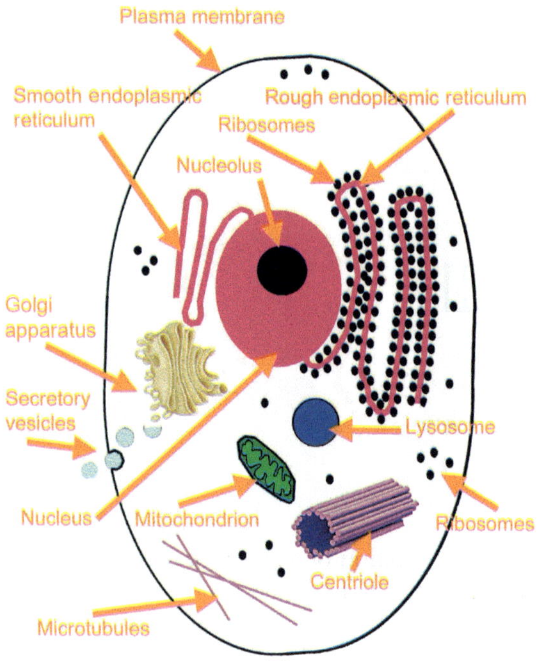

FIGURE 1.1.1 Representative animal cell and its organelles.

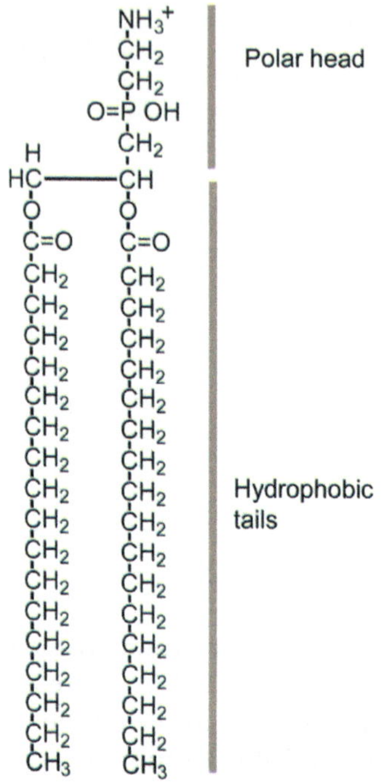

FIGURE 1.1.2 The structure of phospholipids.

and unsaturated carbon, and hydrogen atoms are found in phospholipids in different cell plasma membranes. One possible phospholipid with two stearic acid-derived fatty tails is shown in Figure 1.1.2.

The phospholipids come together to form a bilayer, with the charged phospho groups on the extreme inside and outside of the membrane and the hydrophobic or noncharged fatty tails in the middle of the bilayer membrane. An example of a possible membrane is shown in Figure 1.1.3. The membrane is spanned by receptor proteins that bind a hormone or autocrine on the outside of the cell, and then transmit a protein signal to the inside of the cell. The bilayer is also crossed by channel protein, another type of protein that lets specific molecules, such as ions like Na^+ enter, into the cell (Figure 1.1.3). The charged lipid bilayer blocks proteins, except proteins built into the plasma membrane, and most other molecules from crossing into the cell, except lipids or fats that merge with the plasma membrane lipid bilayers and cross the bilayer. Cells can export proteins across the lipid bilayer in secretory vesicles. Vesicles merge with the membrane after transporting proteins out of the membrane boundaries (see Figure 1.1.1).

NUCLEUS AND NUCLEOLUS

The nucleolus is the most prominent structure within the nucleus. It is the site of ribosomal RNA transcription and processing and ribosome assembly, or the creation of ribosomal protein synthesis centers. Actively growing human cells contain 5-10 million ribosomes that must be copied each time the cell divides into two. The nucleolar organizer regions of chromosomes,

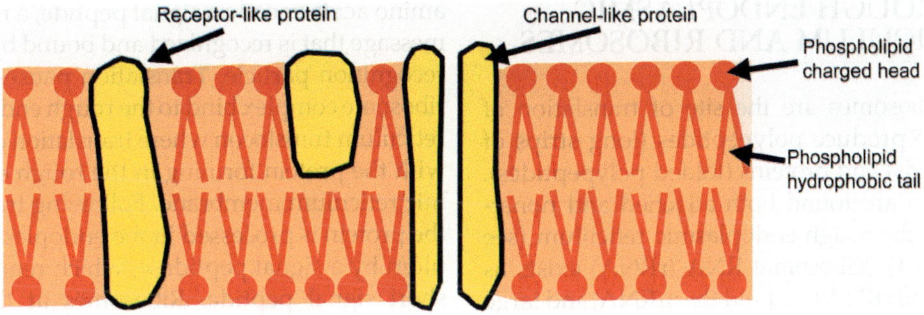

FIGURE 1.1.3 The lipid bilayer structure of plasma membranes.

which contain the genes for ribosomal ribonucleic acid (rRNA), serve as the substance for nucleolar structure. Ribosomal assembly in the nucleolus begins with the transcription of pre-rRNA. During transcription, ribosomal and nonribosomal proteins attach to the RNA. There is modification and cleavage of pre-rRNA and incorporation of more ribosomal proteins and 5S rRNA into maturing preribosomal complexes.

As shown in Figure 1.1.4, the nucleus is a ball filled with chromatin, a DNA-protein complex, with the nucleolus, and multiple loose chromosomes. Each human cell contains 23 pairs of chromosomes or in 46 total. The nucleus is the site where DNA is replicated, and DNA is transcribed to make messenger RNA (mRNA) signals to start the process of translation or protein synthesis which occurs in ribosomes in the cell cytoplasm (all the cell body except the nucleus).

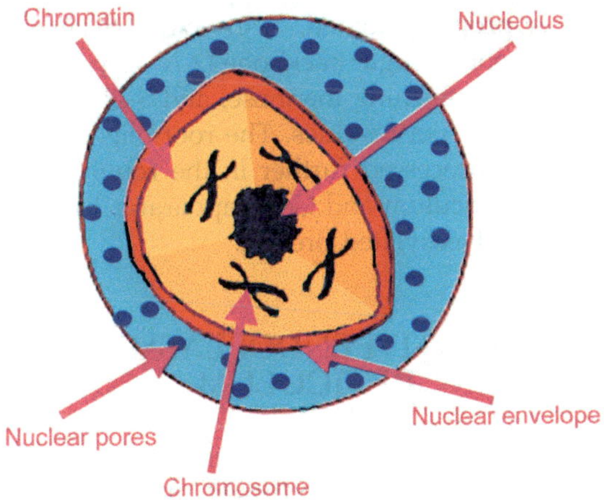

FIGURE 1.1.4 The nucleus and nucleolus.

The function of the nucleus is to maintain the integrity of the genes on chromosomes and to control the activities of the cell by regulating gene expression through mRNA signals. The nucleus is the control center of the cell. Because the nuclear membrane is impermeable to large molecules, nuclear pores are required that regulate nuclear transport of molecules across the envelope. The pores cross both nuclear membranes, providing a channel through which larger molecules must be actively transported by carrier proteins while allowing free movement of small molecules and ions into the nucleus. Movement of large molecules like proteins and mRNA through pores is needed for gene expression and the maintenance of chromosomes.

The nuclear envelope contains thousands of nuclear pores (Figure 1.1.4). Nuclear pores are large protein complexes that cross the nuclear envelope, which is the double membrane surrounding the cell nucleus. The proteins that make up the nuclear pore complex are known as nucleoporins. Each nuclear pore complex is composed of approximately 30 distinct proteins.

ROUGH ENDOPLASMIC RETICULUM AND RIBOSOMES

The ribosomes are the site of translation of mRNA to produce polypeptides (long strips of amino acids) and proteins (folded polypeptides). Ribosomes are found both attached and nonattached to the rough endoplasmic reticulum (see Figure 1.1.1). Ribosomal RNA (rRNA) exists as 30S or small rRNA that hold the mRNA, and large 50S rRNA where protein synthesis occurs (see Figure 1.1.5). As illustrated in Figure 1.1.5, the mRNA binds amino acid-specific transfer RNA (tRNA) that have the corresponding RNA anticodon sequence. The tRNA carries the specific amino acids and places them together in sequence.

The surface of the rough endoplasmic reticulum is studded with ribosomes and has a rough appearance when viewed under an electron microscope. The binding site of the ribosome on the rough endoplasmic reticulum is the translocon. However, the ribosomes are constantly being bound by and released from the rough endoplasmic reticulum. A ribosome only binds to the rough endoplasmic reticulum once a protein-nucleic acid complex forms in the cytoplasm. This special complex forms when a free ribosome begins translating the mRNA of a protein destined for a secretory pathway. The first 5-30 amino acids encode a signal peptide, a molecular message that is recognized and bound by a signal recognition particle. Translation pauses and the ribosome complex bind to the rough endoplasmic reticulum translocon where translation continues with the protein forming in the rough endoplasmic reticulum membrane. Following translation, the protein is processed in the endoplasmic reticulum by a signal peptidase, which removes that short signal peptide. Ribosomes at this point may be released from the rough endoplasmic reticulum back into the cytosol.

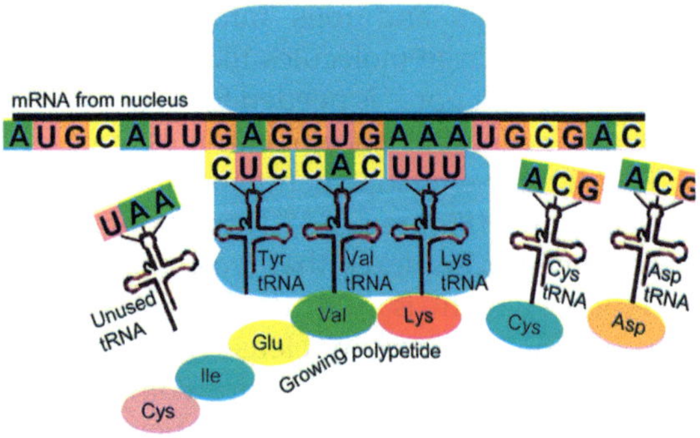

FIGURE 1.1.5 Translation of mRNA into polypeptide sequence occurring in ribosomes.

The membrane of the rough endoplasmic reticulum links to the outer layer of the nuclear envelope (Figure 1.1.1). While there is no continuous membrane between the endoplasmic reticulum and the Golgi apparatus, transport vesicles shuttle translated proteins between these two compartments. Transport vesicles are surrounded by coating proteins called COPII, which target vesicles to move proteins to the Golgi apparatus to complete processing.

The rough endoplasmic reticulum has multiple functions. These include manufacture of enzymes and secreted proteins. Glycosylation is the process whereby sugars are added to proteins as side chains. Initial glycosylation of proteins, as in N-linked or Asn-linked glycosylation, occurs in the rough endoplasmic reticulum. Glycosylation continues in the Golgi apparatus. If the protein is properly folded, a glycosyltransferases adds a root 14-sugar backbone (2-*N*-acetylglucosamine, 9-branching mannose, and 3-glucose at the end) to Asn residues. The root oligosaccharides are processed further in the rough endoplasmic reticulum and Golgi apparatus to make the final glycosylated protein.

SMOOTH ENDOPLASMIC RETICULUM

The smooth endoplasmic reticulum functions in many metabolic processes. It synthesizes lipids, phospholipids as in plasma membranes, and steroids. Cells that secrete these products, such as cells of the testes, ovaries, and skin oil glands, have an excess of smooth endoplasmic reticulum. The smooth endoplasmic reticulum also carries out the metabolism of carbohydrates and steroids. In muscle cells, the smooth endoplasmic reticulum regulates calcium ion storage. The smooth endoplasmic reticulum like the rough endoplasmic reticulum is connected to the nuclear envelope. The smooth endoplasmic reticulum comprises tube-like structure located near the cell periphery. These tubules or tubes sometimes branch forming a network that is reticular in appearance. The network of smooth endoplasmic reticulum allows for an increased surface area to be devoted to storage of key enzymes.

GOLGI APPARATUS

The Golgi apparatus is an organelle found in all animal cells (Figure 1.1.6). It was first identified in 1897 by the Italian physician Camillo Golgi and thereafter named after him. The Golgi apparatus plays an important role in the synthesis of proteoglycans, mixtures of amino acids and sugars, which are molecules present in the extracellular matrix of animals. It is also a major site of carbohydrate synthesis or glycosylation of proteins. Proteins are *O*-glycosylated in the Golgi apparatus and *N*-glycosylation is completed in this organelle. This includes the production of glycosaminoglycans, long unbranched polysaccharides the Golgi then attaches to a protein synthesized in the endoplasmic reticulum to form proteoglycans.

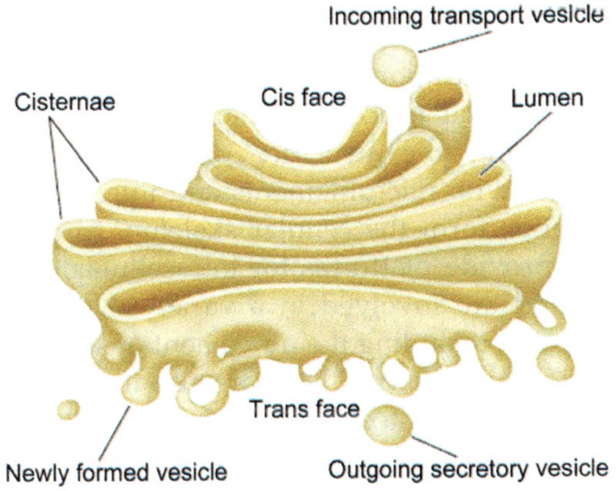

FIGURE 1.1.6 The Golgi apparatus.

Another task of the Golgi involves the sulfation of certain molecules passing through its lumen via sulfotranferases that gain their sulfur molecule from a donor called PAPS (3′-phosphoadenosine-5′-phosphosulfate). This process occurs on the glycosaminoglycans of proteoglycans as well as on the core protein. Sulfation is generally performed in the trans-Golgi network. The level of sulfation is very important to proteoglycan signaling abilities as well as giving the proteoglycan its overall negative charge.

Transport vesicles carry proteins from the rough endoplasmic reticulum to the cis face of the Golgi apparatus, where they fuse with the Golgi membrane and empty their contents into the Golgi lumen. Once inside the lumen, the molecules are modified, glycosylated, or sulfated, then sorted for transport to their next destinations.

LYSOSOMES

The lysosomes are spherical vesicles each containing hydrolytic enzymes, which are capable of breaking down or degrading virtually all kinds of biomolecules, including proteins, nucleic acids, carbohydrates, lipids, and cellular debris. Lysosomes contain more than 50 different enzymes, which are all active at an acidic environment of about pH 5.0. Thus, they act as garbage dump system of the cell by digesting unwanted materials, whether proteins, carbohydrate, nucleic acid, or other cell debris in the cytoplasm.

Enzymes of the lysosomes are made on the ribosomes of the rough endoplasmic reticulum.

The enzymes are released from Golgi apparatuses in small vesicles that ultimately fuse with lysosomal acidic vesicles. Extracellular materials such as microorganisms taken up by phagocytosis, macromolecules by endocytosis, and unwanted cell organelles are fused with lysosomes in which they are broken down to their basic molecules. Thus, lysosomes are the recycling units of a cell. Synthesis of lysosome enzymes are controlled by nuclear genes. Mutations in the genes for these enzymes are responsible for more than 30 different human genetic diseases, which are known as lysosomal storage diseases.

MITOCHONDRION

Mitochondrion is considered to be the power stations of animal cells, in that this is the site of production of adenosine triphosphate (ATP), the power source of all animal energy. ATP is cleaved to form adenosine diphosphate (ADP) to release the bond energy. The ATP power source generation is driven by a remarkable ionic process called oxidative phosphorylation, which is driven by the oxygen we breathe and NADH and $FADH_2$ from sugar degradation. Chapter 1.2 examines oxidative phosphorylation. Oxidative phosphorylation in mitochondria follows the equation of life, $1\text{Glucose} + 6O_2 + 38\text{ADP} \rightarrow 6CO_2 + 6H_2O + 38\text{ATP}$.

The number of mitochondria in a cell varies widely by organism and tissue type. Many cells have only a single mitochondrion, whereas others can contain several thousand mitochondria, depending on the energy needs of the cell. Mitochondria are composed of compartments that carry out specialized functions, and include the outer membrane, the intermembrane space, the inner membrane, and the cristae and matrix.

Although most of a cell's DNA is contained in the cell nucleus, the mitochondrion has its own independent genome stored in a circular DNA loop in the mitochondrion. Mitochondrial DNA codes for mitochondrial proteins. Oddly, mitochondria and mitochondrial DNA is strictly inherited through the reproduction pathways from the mother only.

Many enzymatic pathways occur in the mitochondrion, generally those involving ATP. The citric acid loop or Krebs cycle, the key pathway of acetyl-CoA usage and conversion of sugars to CO_2, and generation of energy (NADH and $FADH_2$) occurs inside the mitochondrion, as does fatty acid ß-oxidation, a pathway that generates much ATP.

The concentrations of free calcium in the cell can regulate an array of reactions and is important for signal transduction in the cell. Mitochondria can transiently store calcium in a cell. Mitochondrion also controls cell apoptosis or cell death in a cell. An ion channel is formed on the outer mitochondrial membrane in response to certain apoptotic stimuli (see Chapter 2.3 Apoptosis, Growth, and Aging).

MICROTUBULES AND CENTRIOLES

Microtubules are a part of the cellular skeleton, found throughout the cell. These tubular polymers are made of the protein tubulin and grow an average length of 25 μm. The outer diameter of a microtubule is about 24 nm, while the inner diameter is about 12 nm. They are found in all animal cells and are formed by dimers of two globular proteins, α and ß-tubulin. Microtubules are important in a number of cellular processes. They are involved in maintaining the structure of the cell and, together with microfilaments and intermediate filaments, form the cellular skeleton. Microtubules provide platforms for intracellular transport and are involved in a variety of cellular processes, including the movement of secretory vesicles within a cell. They are also involved in cell division (mitosis and meiosis), including the formation of mitotic spindles, which are used to pull apart chromosomes.

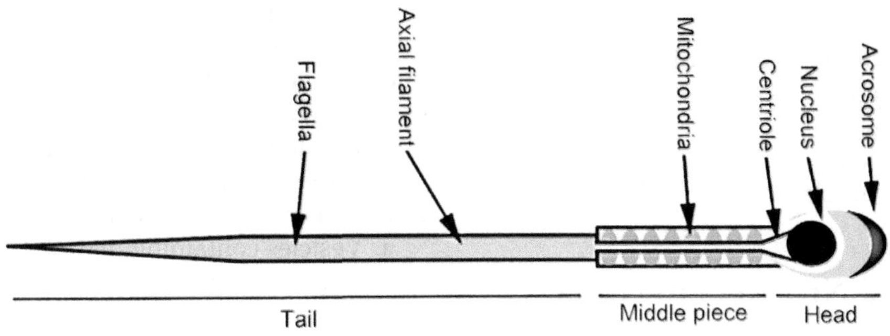

FIGURE 1.1.7 Spermatozoon.

Centrioles are made from tubulin or microtubules. They form cylindrical structure from nine sets of microtubule triplets, arranged in a cylinder. Centrioles are involved in the physical division of cells, in the organization of the mitotic spindle.

CELLS, CELLS, AND MORE CELLS

Cells have hundreds of shapes in the human body serving different purposes and functions. Looking, for instance at a spermatozoon or mature sperm cell, the nucleus is in the head of the spermatozoon and the mitochondrion or power source is all together in the middle piece of the spermatozoon to provide power axial filament movements (Figure 1.1.7).

The erythrocyte or red blood cell is a cell with no nucleus at all and therefore is missing most other cellular organelles. All the space is used for storing the protein hemoglobin. The erythrocyte cannot divide or make itself and is made by bone marrow cells. The erythrocyte in humans is the principal means of delivering oxygen to the body tissues via the blood flow through the circulatory system. Many extreme cells are found in humans all basically built around the structure shown in Figure 1.1.1. Spermatozoon and erythrocytes are examples of extreme cells.

CHAPTER KEYWORDS

Know these keywords and what each word refers to as well as its function. Read over the list. Are all these keywords familiar?

Bilayer
Centrioles
Chromatin
Cis-Golgi
Cytoplasm
COPII coating proteins
Erythrocyte
Golgi apparatus
Golgi cisternae
Golgi lumen
Lysosome
Messenger RNA
Microtubules
Mitochondrion
Nuclear pores
Nuclear envelope
Nucleus
Nucleolus
Plasma membrane
Phospholipid
Ribosome
Ribosomal RNA
Root 14 sugar backbone
Rough endoplasmic reticulum

Secretory vesicle
Smooth endoplasmic reticulum
Spermatozoon
Trans-Golgi
Transfer RNA
Transport vesicle

INTERNET REFERENCES

Use these references to find out more.
en.wikipedia.org/wiki/Nucleolus
en.wikipedia.org/wiki/Cell_nucleus
en.wikipedia.org/wiki/Endoplasmic_reticulum
en.wikipedia.org/wiki/Ribosome
en.wikipedia.org/wiki/Golgi_apparatus
en.wikipedia.org/wiki/Mitochondrion
en.wikipedia.org/wiki/Microtubule
en.wikipedia.org/wiki/Centriole
en.wikipedia.org/wiki/Spermatozoon
en.wikipedia.org/wiki/Red_blood_cell

CHAPTER QUIZ

Answer these questions. Do you know the answers, holy cow! Check yourself out in this quiz, and prove that you are a humanology wiz.

1. The nucleolus is located where and does what?
   - **a.** Located on smooth endoplasmic reticulum, site of mRNA translation
   - **b.** Located in lysosome, responsible for transport within cell
   - **c.** Located within the nucleus, site of ribosome assembly.
   - **d.** Located in plasma membrane, responsible for receptor synthesis
   - **e.** Located adjacent to plasma membrane, processes ribosomal RNA
2. What is a lysosome, what is its function?
   - **a.** Vesicles containing hydrolytic enzymes to degrade molecules
   - **b.** Part of plasma membrane, permits anions to assess cells
   - **c.** Vesicles that control transport of molecules within cells
   - **d.** Site of production of ribosomes
   - **e.** Site of production of ribosomal RNA
3. Mitochondrial cytochromes are coded for by:
   - **a.** Chromosomes in the nucleus
   - **b.** Mitochondrial circular DNA
   - **c.** By mRNA on the endoplasmic reticulum
   - **d.** By lysosomal DNA
   - **e.** None of the above
4. The mitochondrion of cells is the site of:
   - **a.** Glycolysis and the citric acid cycle
   - **b.** Oxidative phosphorylation and apoptosis regulation
   - **c.** Apoptosis regulation and fatty acid synthesis
   - **d.** Protein synthesis and fatty acid degradation
   - **e.** Glycolysis and gluconeogenesis
5. The plasma membrane contains:
   - **a.** Phospholipids and imbedded sugars
   - **b.** Phospholipids and imbedded proteins
   - **c.** DNA and microzymes
   - **d.** Imbedded proteins and sugars
   - **e.** Imbedded proteins and DNA

CHAPTER

1.2

Human Chemistry

Humans, like other animals, are filled with a wide range of organic molecules. If you are unfamiliar with these molecules, it may be wise to study some of the organic chemistry groups found in basic human molecules (Table 1.2.1).

Human molecules primarily fall into four classes: sugars, fatty acids, amino acids, and nucleotides. Each of these has a characteristic basic unit structure, and a classical polymer structure. Polymers (units merge by condensing H^+ and OH^- groups) involving three units to 500 units linked back to back. Figure 1.2.1 shows the structure of common sugars, fatty acids, amino acids, and nucleotides. Polymers are formed of amino acids, which are called proteins. Polymers are formed of sugars called starch and glycogen. Polymers are formed of nucleotides called DNA and RNA. Fatty acids link together in unusual ways forming triglycerides, large folded fatty acids such as steroids and prostaglandins and solid mazes of fatty acids such as membranes. Figure 1.2.2 shows how these molecules condense together (DNA translation, glycogenesis, and DNA synthesis involve complex pathways). Chapters 1.3–1.5 describe these pathways.

Examples of fatty acid combinations and folding derivatives are shown in Figure 1.2.3. These include triglycerides, common fatty acid derivatives in blood, and steroids and prostaglandins or folded fatty acids.

Polymers of amino acid, sugars, and nucleotides can be lined up together. For example, glycoproteins are blends of amino acid polymers or proteins and sugar polymers as side chains. Immunoglobulins and mucins are glycoproteins, as are many protein hormones, membrane proteins, and structural glycoproteins in connective tissue. The glycoprotein human chorionic gonadotropin (hCG), its bivalent peptide (two-stranded) amino acid sequence, its three-dimensional structure (blue is α-subunit and red is ß-subunit), and its subunit DNA sequence are all shown in Figure 1.2.4.

Other sugar-protein combinations include proteoglycans or glycosaminoglycans found in connective tissue. Combinations of lipids or fatty acids and sugars form gycolipids. These are mostly found in membranes. Combinations of nucleotides and proteins also exist, which are called nucleopeptides. The most well-known human nucleopeptide is the enzyme tyrosyl-DNA-phosphodiesterase-1.

http://dx.doi.org/10.1016/B978-0-12-803699-0.00016-5

TABLE 1.2.1 Organic chemistry groups and structures found in many human molecules.

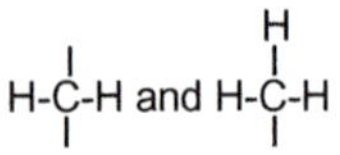	Hydrocarbon and methyl groups as found in fatty acids and many other molecules
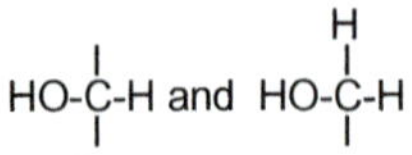	Hydroxyl-carbon group as found in sugars and other molecules
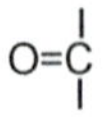	Acyl groups, keto- or aldehyde groups found in assorted keto and aldehyde molecules
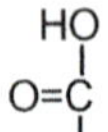	Carboxylic acid group, found in amino acids, fatty acids, and other acidic molecules. Negatively charged COO^-
	Amino group found in amino acids, positively charges NH_3^+
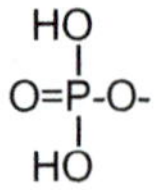	Phosphate group found in nucleotides and other molecules, negatively charged $H_2PO_4^-$
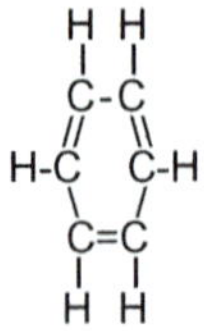	Aromatic ring structure

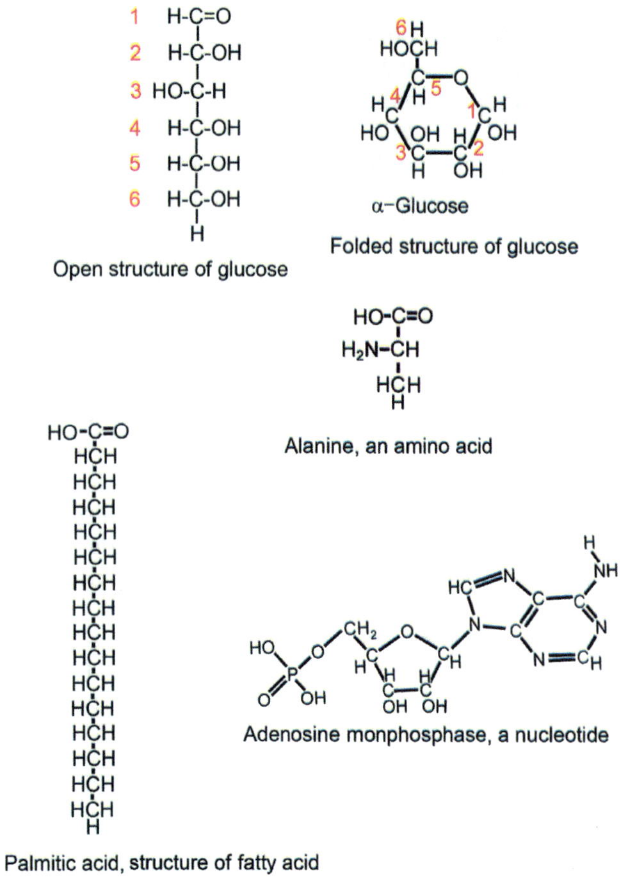

FIGURE 1.2.1 Structure of sugar, fatty acid, amino acid, and nucleotide.

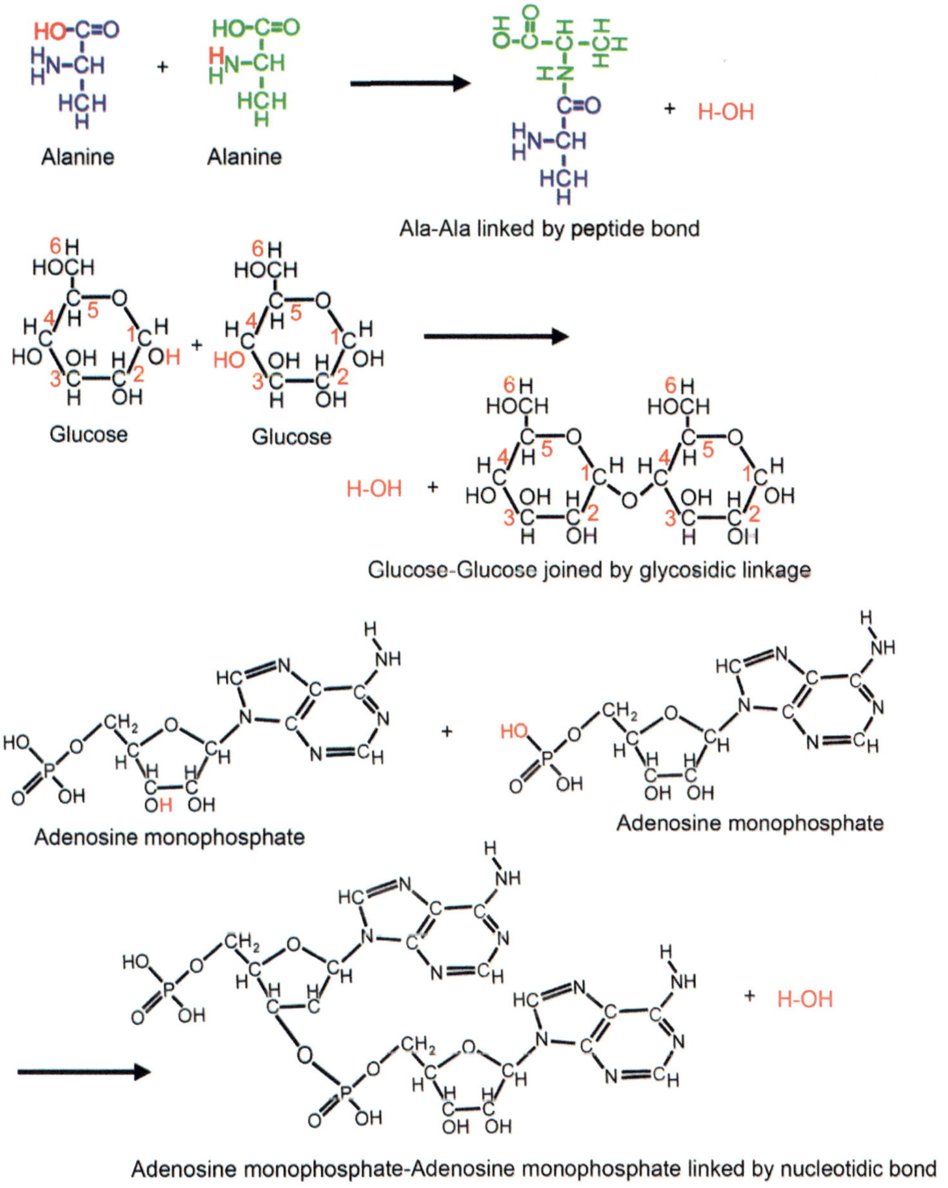

FIGURE 1.2.2 Amino acids forming peptide dimers, sugars forming glycosidic dimers, and nucleotides forming nucleotidic dimers.

Estradiol as an example of folded fatty acid or steroid

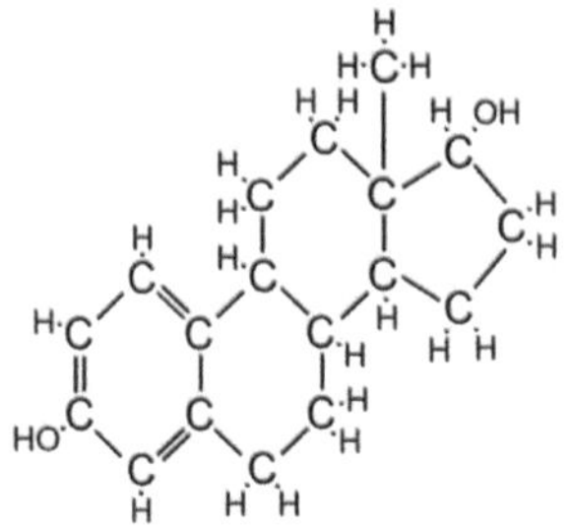

Triglyceride as an example of multi-fatty acid

Prostaglandin E2 as an example of folded fatty acid

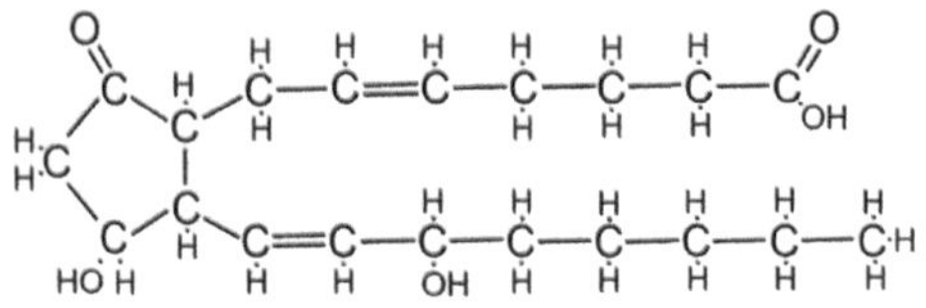

FIGURE 1.2.3 Fatty acid combinations and folding derivatives.

CHAPTER KEYWORDS

Know these keywords and what each word refers to as well as its function. Read over the list. Are all these keywords familiar?

Fatty acid
Glycolipid
Glycoprotein
Glycosaminoglycans
Glycosidic
Nucleopeptide
Nucleotide
Nucleotidic
Peptide
Prostaglandin
Protein
Proteoglycan
Steroid
Sugar

ala-pro-asp-val-gln-asp-cys-pro-glu-cys-thr-leu-gln-glu-asp-pro-phe-phe-ser-gln-pro-gly-ala-pro-ile-leu-gln-cys-met-gly-cys-cys-phe-ser-arg-ala-tyr-pro-thr-pro- leu-arg-ser-lys-lys-thr-met-leu-val-gln-lys-asn-val-thr-ser-glu-ser-thr-cys-cys-val-ala-lys-ser-tyr-asn-arg-val-thr-val-met-gly-gly-phe-lys-val-glu-asn-his-thr-ala-cys-his-cys-ser-thr-cys-tyr-tyr-his-lys-ser

ser-lys-glu-pro-leu-arg-pro-arg-cys-arg-pro-ile-asn-ala-thr-leu-ala-val-glu-lys-glu-gly-cys-pro-val-cys-ile-thr-val-asn-thr-thr-ile-cys-ala-gly-tyr-cys-pro-thr-met-thr-arg-val-leu-gln-gly-val-leu-pro-ala-leu-pro-gln-val-val-cys-asn-tyr-arg-asp-val-arg-phe-glu-ser-ile-arg-leu-pro-gly-cys-pro-arg-gly-val-asn-pro-val-val-ser-tyr-ala-val-ala-leu-ser-cys-gln-cys¬-ala-leu-cys-arg-arg-ser-thr-thr-asp-cys-gly-gly-pro-lys-asp-his-pro-leu-thr-cys-asp-asp-pro-arg-phe-gln-asp-ser-ser-ser-ser-lys-ala-pro-pro-pro-ser-leu-pro-ser-pro-ser-arg-leu-pro-gly-pro-ser-asp-thr-pro-ile-leu-pro-gln

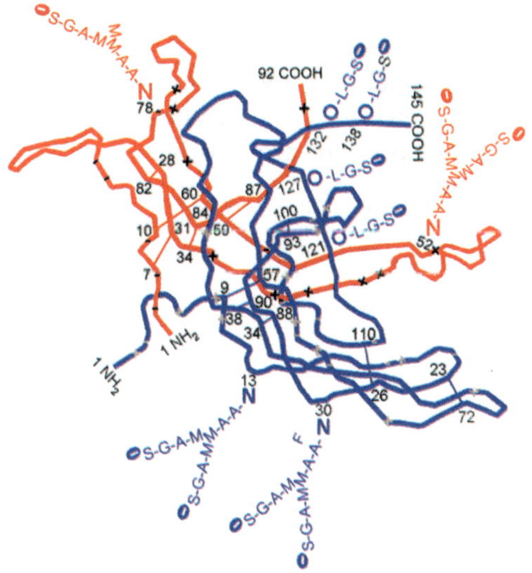

In amino acid sequence (top), three-dimensional structure (middle), and DNA sequence (bottom) α-subunit is shown in red and β-subunit is shown in blue.

DNA sequence is shown in single letter code, A is deoxyadenosine-monophosphate, G is deoxy-guanosine monophosphate, C is deoxycytidine monophosphase and T is deoxythymidine mono-phosphate.

In the three-dimensional structure sugars are indicated by single letter code, G is galactose A is *N*-acetylglucosamine, M is mannose, F is Fucose, S is sialic acid and L is *N*-acetyl-galactosamine.

ATGGATTACTACAGAAAATATGCAGCTATCTTTCTGGT
CACATTGTCGGTGTTTCTGCATGTTCTCCATTCCGCTCCTGATGTGCAGGAGACAGGGTT
TCACCATGTTGCCCAGGCTGCTCTCAAACTCCTGAGCTCAAGCAATCCACCCACTAAGGC
CTCCCAAAGTGCTAGGATTACAGATTGCCCAGAATGCACGCTACAGGAAAACCCATTCTT
CTCCCAGCCGGGTGCCCCAATACTTCAGTGCATGGGCTGCTGCTTCTCTAGAGCATATCC
CACTCCACTAAGGTCCAAGAAGACGATGTTGGTCCAAAAGAACGTCACCTCAGAGTCCAC
TTGCTGTGTAGCTAAATCATATAACAGGGTCACAGTAATGGGGGGTTTCAAAGTGGAGAA
CCACACGGCGTGCCACTGCAGTACTTGTTATTATCACAAATCTTAA

ATGGAGATGTTCCAGGGGCTGCTGCTGTTGCTGCTGCTGAGCATGGGCGGGACATGGGCA
TCCAAGGAGCCGCTTCGGCCACGGTGCCGCCCCATCAATGCCACCCTGGCTGTGGAGAAG
GAGGGCTGCCCCGTGTGCATCACCGTCAACACCACCATCTGTGCCGGCTACTGCCCCACC
ATGACCCGCGTGCTGCAGGGGGTCCTGCCGGCCCTGCCTCAGGTGGTGTGCAACTACCGC
GATGTGCGCTTCGAGTCCATCCGGCTCCCTGGCTGCCCGCGCGGCGTGAACCCCGTGGTC
TCCTACGCCGTGGCTCTCAGCTGTCAATGTGCACTCTGCCGCCGCAGCACCACTGACTGC
GGGGGTCCCAAGGACCACCCCTTGACCTGTGATGACCCCCGCTTCCAGGACTCCTCTTCC
TCAAAGGCCCCTCCCCCCAGCCTTCCAAGCCCATCCCGACTCCCGGGGCCCTCGGACACC
CCGATCCTCCCACAATAA

FIGURE 1.2.4 The amino acid sequence, three-dimensional structure, and DNA sequence of hCG.

INTERNET REFERENCES

Use these references to find out more.
en.wikipedia.org/wiki/Amino_acid
en.wikipedia.org/wiki/Sugar
en.wikipedia.org/wiki/lipid
en.wikipedia.org/wiki/glycoprotein
en.wikipedia.org/wiki/Glycolipid
en.wikipedia.org/wiki/nucleotide

CHAPTER QUIZ

1. Answer these questions. Do you know the answers, holy cow! Check yourself out in this quiz, and prove that you are a humanology wiz.
2. A prostaglandin is a
   a. Mixture of fatty acids and amino acids
   b. Folded fatty acid
   c. Mixture of sugars and amino acids
   d. Mixture of nucleotides and sugars
   e. Made up of amino acids
3. Glycoproteins are a mixture of sugar polymers and amino acid polymers. Which statement is true?
   a. Multiple hormone are glycoproteins
   b. Cell membrane are made of primarily glycoproteins
   c. Hemoglobin is a glycoprotein
   d. Most enzymes are glycoproteins
   e. Blood proteins are mostly glycoproteins

CHAPTER

1.3

Sugars, Fatty Acids, and Energy Biochemistry

This chapter describes the basic biochemistry of humans, specifically how humans use sugars and fats as energy and how it actually functions to power muscle movements, power the heart, power nerve signals, and power biochemical reactions. In many respects, life as we know it is these amazing energy reactions. These are the reactions of life. A dead cell does not generate or use energy, and a living cell relies on its energy sources. Here we describe how energy is generated by living cells and how it is used.

Figure 1.3.1 outlines the different biochemical pathways that run human and animal energetics, provide energy when needed, and use energy for storing vital molecules. The central pathways are "glycolysis," "ß-oxidation," and the "citric acid cycle," or the breaking down of glucose to two molecules of pyruvate, and then using pyruvate and breaking down fats by ß-oxidation to generate reduced nicotinamide adenine dinucleotide (NADH), a derivative of the vitamin niacin, and reduced flavin adenine dinucleotide ($FADH_2$), a derivative of the vitamin riboflavin, which both drive oxidative phosphorylation or energy generation (Figure 1.3.2). Glucose is produced when energy is not needed and stored as glycogen, sucrose, and other saccharides (Figure 1.3.1). Acetyl CoA is stored as fatty acids, major potential sources of NADH, and oxidative phosphorylation.

ATP is the main source of energy in humans, powering the heart, muscles, brain, and nerves. ATP is generated from NADH and $FADH_2$ by oxidative phosphorylation in mitochondria in cells (Figure 1.3.2). In this respect, energy is transferred in the body to reduced NADH and $FADH_2$ in enzymatic reactions. In mitochondria, NADH and $FADH_2$ are oxidized by electron transport reaction to generate ATP through oxidative phosphorylation (Figure 1.3.2). In this respect, they are both electron-transport mechanisms and energy conveyance molecules. Oxidative phosphorylation requires oxygen, which is why we breathe. Examining glycolysis of glucose, citric acid cycle, and oxidative phosphorylation together the equation is $1\text{Glucose} + 6O_2 + 38\,\text{ADP} \rightarrow 6CO_2 + 6H_2O + 38\,\text{ATP}$

Oxidative phosphorylation in the mitochondrion is in many respects the human or animal power supply. Through electron transport in the

Human Physiology, Biochemistry and Basic Medicine
http://dx.doi.org/10.1016/B978-0-12-803699-0.00019-0

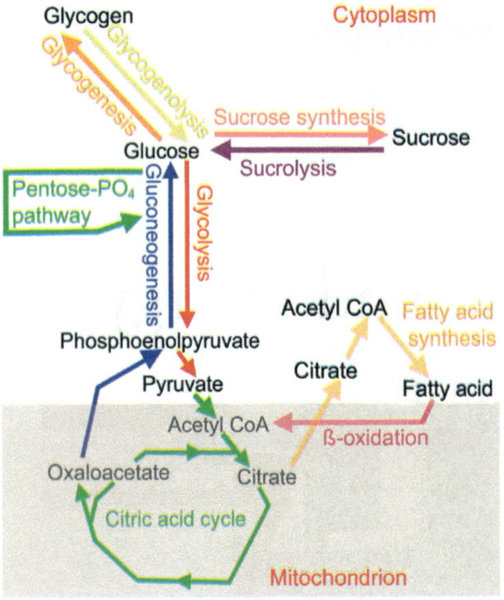

FIGURE 1.3.1 Overview of carbohydrate and fatty acid biochemistry.

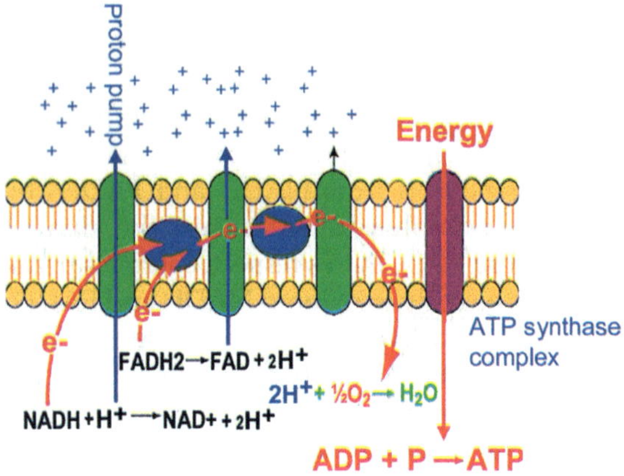

FIGURE 1.3.2 Oxidative phosphorylation in mitochondria.

membrane of mitochondria the energy of NADH and the energy of $FADH_2$ are used along with oxygen to power ATP synthesis (Figure 1.3.2). Each NADH molecule powers the generation of three ATP molecules and each $FADH_2$ molecule powers a shorter electron transport cycle or the generation of two ATP molecules.

Each molecule of glucose processed to CO_2 generates 4 ATP molecules in glycolysis and uses 2 ATP molecules (net 2 ATP molecules) through the glycolysis pathway. It generates 2 further molecules of guanosine triphosphate (GTP) in the citric acid cycle. This GTP is converted to ATP by nucleoside-diphosphate kinase. As such, a balance of 2+2 or 4 ATP molecules is made with the processing of each glucose molecule. The biochemical pathway generates 10 NADH molecules (the electrons for generating 30 ATP molecules in oxidative phosphorylation) and 2 $FADH_2$ molecules (the electrons for generating 4 ATP molecules). Summing everything together, each molecule of glucose provides the energy needed to generate 4+30+4 or potentially 38 ATP molecules.

The central pathways for powering glycolysis or breakdown of glucose, gluconeogenesis, or glucose generation, glycogenolysis and glycogenesis or synthesis and breakdown of glycogen, fatty acid synthesis and fatty acid ß-oxidation are shown in Figure 1.3.1. All these pathways are very much regulated by hormones and other feedback mechanism so that the body's energy needs and storage needs can match the body's food input. Insulin, which rids the blood of excess glucose, promotes glycolysis, glycogenesis, and fatty acid synthesis. Glucagon and antiinsulin hormone made by the pancreas does the reverse, promoting glycogenolysis and inhibiting glycolysis leading to an accumulation of glucose. Epinephrine or adrenaline promotes glycogenolysis or breakdown of stored sugars throughout the body, glucose accumulation or gluconeogenesis in the liver blocking glycolysis, but does not block muscle glycolysis or muscle energy generation from glucose. Epinephrine accumulates body glucose for use in glycolysis by muscles in times of exercise.

All hormones work by activating or phosphorylating key enzymes. The key enzymes in glycolysis and gluconeogenesis are hexokinase, hosphofructokinase, and pyruvate dehydrogenase. In fatty acid metabolism, the key enzymes are lipase (fatty acid degradation), fatty acid synthetase, and acetyl CoA carboxylase (fatty acid synthesis). In glycogen formation, the key enzymes are glycogen synthase (synthesis) and glycogen phosphorylase (breakdown). These are the enzymes that are directly modulated by hormones.

Let us start by looking at the top of the ladder at glycogen metabolism. It is noteworthy, before we start, to point out that some sugar and fat processing enzymes are located in cell mitochondrion, while other are located in cell cytoplasm (Figure 1.3.1). Glycogen synthesis and breakdown enzymes are all located in the cytoplasm.

Glycogen is a large branched polymer of glucose that serves as a form of energy storage in animals. In humans, glycogen is made and stored in liver and muscle cells. Muscle cell glycogen is broken down into glucose, and liver glycogen is broken down into glucose as a circulating energy source glucose for use by the body. Glycogen is an energy reserve that can be quickly mobilized to meet sudden needs. In liver cells, glycogen can comprise up to 8% of liver weight. In muscles, glycogen is found at lower concentrations, 2% of weight. Small amounts of glycogen are found in the kidneys and lesser amounts in glial cells in the brain. The uterus stores glycogen during pregnancy as a reserve to nourish the embryo.

As shown in Figure 1.3.3, glycogen synthesis in the liver and muscles involves phosphorylation of glucose to glucose-6-phosphate and then glucose-1-phosphate, the addition of a nucleotide, a uridine diphosphate group to glucose (uridine phosphate-glucose), and then linkage of glucose molecules back to back with release of the uridine diphosphate group. Figure 1.3.3 shows a α-glucose molecule with carbon ring numbering shown in red. Carbon 1 is the primary carbon. If the hydroxyl group attached to carbon is under the carbon as shown, it is an α-glucose molecule. If it is over the glucose, it is a ß-glucose molecule.

Multiple isomers of sugars exist with identical compositions. α-galactose, for instance, only differs from α-glucose by a reversal of the hydroxyl group attached to Carbon 4, with the hydroxyl group above the carbon instead of underneath. Galactose uses the same numbering system as glucose. Very similar synthesis pathways to glycogen using galactose-1-phosphate molecules and uridine diphosphate-galactose intermediates make sugar dimers, trimers, tetramers, and smaller polymers. Sucrose or sweetening sugar as made in plants is a dimer comprising glucose and the sugar fructose, or α-glucose 1,2 linked to ß- fructose. Lactose or human milk sugar is made in humans. It is ß-galactose 1,4 linked to glucose. Maltose is α-glucose 1,6 linked glucose. The α and ß 2, 4, and 6 linkages in these disaccharides, versus the α-subunit 1,4 linkages in glycogen shows the variability of how sugars are linked.

While plant sucrose is a disaccharide (two sugars), it is specifically cleaved in humans by the enzyme sucrase into free fructose and free glucose, and the enzyme lactase in humans specifically cleaves lactose a disaccharide in human or animal milk sugar into galactose in glucose. A more complicated hydrolysis protocol is required for splitting liver and muscle glucose polymers or glycogens. As shown in Figure 1.3.3, glycogen is lysed by glycogen phosphorylase to glucose-1-phosphate, which phosphorylates sugars. This is converted by phosphoglucomutase to glucose-6-phosphate, which is either used for glycolysis or generation of NADH or the energy carrier ATP, or converted into glucose.

Following the breakdown of glycogen, sucrose, and lactose into glucose-6-phosphate and glucose, saccharides are used for generating energy through the key glycolysis pathways and citric acid cycle. The glycolysis pathway is shown in Figure 1.3.4. The end of this pathway is pyruvic acid, which is transported into the mitochondrion and converted into acetyl-CoA for use in the citric acid cycle. The reverse pathway is gluconeogenesis or synthesis of glucose.

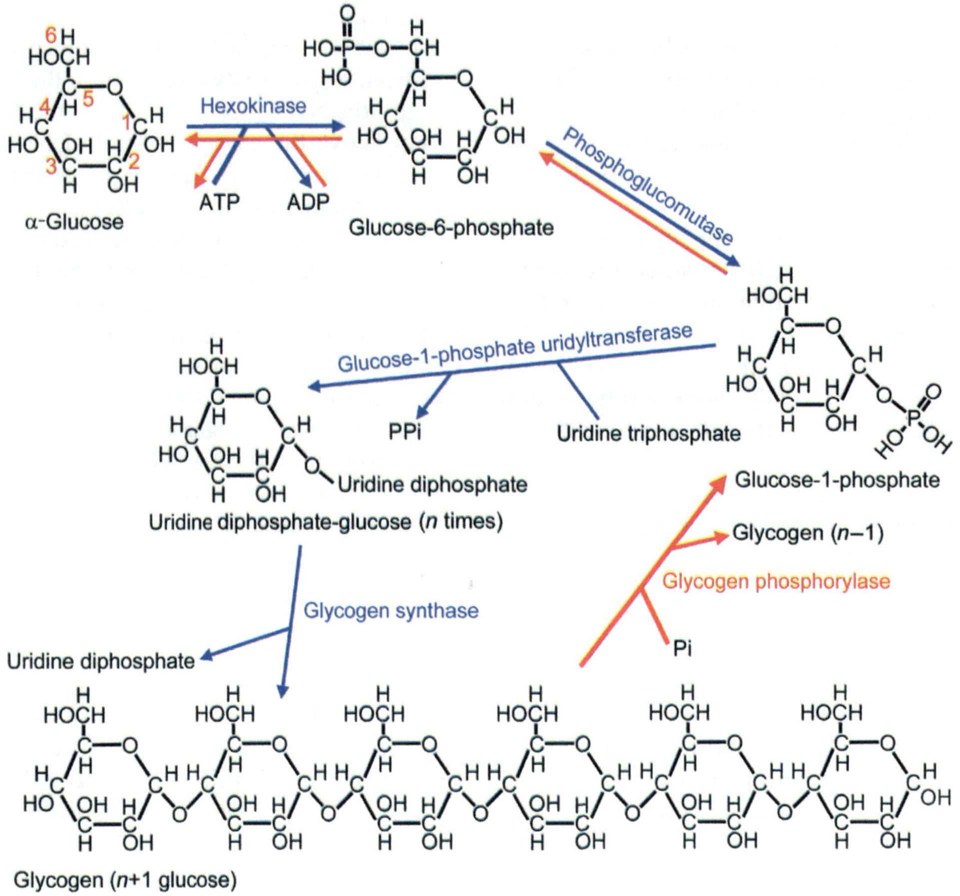

FIGURE 1.3.3 Pathways of glycogen synthesis (blue lines) and glycogenolysis (red lines) in human liver and muscles.

This pathway starts with oxaloacetic acid from the mitochondrion, which is converted into phosphoenolpyruvate by phosphoenolpyruvate carboxykinase, where it joins a reverse glycolysis pathway (Figure 1.3.4).

During glycolysis, glucose is first phosphorylated to glucose-6-phosphate and then converted to fructose-6-phosphated (same composition). After the addition of a second phosphate, fructose-1,6-biphosphate, this hexose sugar is split into two isomeric triose phosphate sugars, glyceraldehyde phosphate and dihydroxyacetone phosphate. The second isomer, dihydroxyacetone phosphate, is quickly isomerized to the first isomer, glyceraldehyde phosphate. The two molecules of glyceraldehyde phosphate generated are then subject to the first NADH or energy-generating dehydrogenase reaction. The bisphosphoglycerate or acidic triose generated is first refined to 3-phosphoglycerate and then to 2-phosphoglycerate, before being converted to the acidic sugar phosphoenolpyruvate (Figure 1.3.4). Phosphoenolpyruvate then loses its phosphate group to generate an ATP and the acidic sugar pyruvic acid. Pyruvic acid enters the mitochondrion where it is converted to acetyl-CoA.

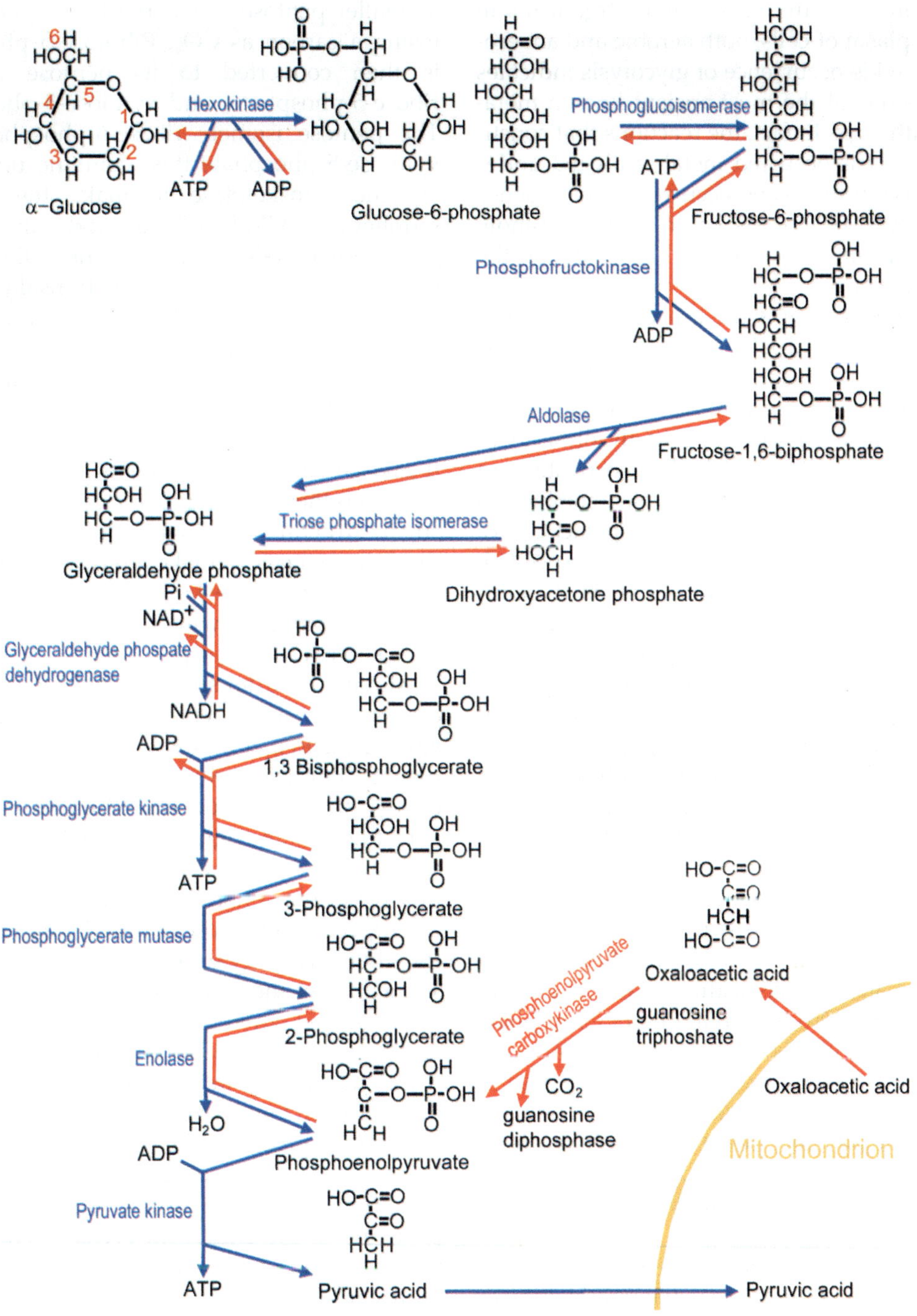

FIGURE 1.3.4 Pathways of glycolysis (blue lines) and of gluconeogenesis (red lines).

Glycolysis occurs in nearly all organisms in the cytoplasm of cells, both aerobic and anaerobic. The wide occurrence of glycolysis indicates that it is one of the most ancient known metabolic pathways. In fact, the reactions that constitute glycolysis occurred metal-catalyzed under Archean conditions and occurred in the absence of enzymes. Glycolysis could have somehow evolved from the chemical restraints of the prebiotic world.

Gluconeogenesis is the exact reverse of glycolysis. Mitochondrial oxaloacetate is released into the cytoplasm and converted by the enzyme phosphoenolpyruvate carboxykinase into phosphoenolpyruvate. It then reverses glycolysis making the trioses, glyceraldehyde phosphate, and dihydroxyacetone phosphate, which combine to form the hexose fructose-1,6-biphosphate. Gluconeogenesis generates blood glucose when needed.

The pentose phosphate shunt is a semi-independent alternative pathway that parallels glycolysis (Figure 1.3.5). It generates the reducing agent reduced nicotinamide adenine dinucleotide phosphate (NADPH), which is independent of the NADH of oxidative phosphorylation, and pentoses. There are two distinct phases in the pentose phosphate shunt. The first is the oxidative phase, in which NADPH is generated, and the second is the synthesis of 5-carbon sugars (Figure 1.3.5). The primary purpose of this alternative pathway is the generation of reducing equivalents, in the form of NADPH, used in reductive biosynthesis within cells, such as fatty acid synthesis. Production of ribose-5-phosphate used in the synthesis of nucleotides or RNA and production of erythrose-4-phosphate are used in the synthesis of aromatic amino acids. In humans and mammals, the pentose phosphate shunt occurs exclusively in the cytoplasm of cells, and is found to be most active in the liver, mammary gland, and adrenal cortex.

The pentose phosphate shunt makes a hexose 6-phosphogluconate intermediate from glucose-6-phosphate. This is enzymatically degraded to a smaller pentose sugar, ribulose-5-phosphate, losing a carbon as CO_2. Ribulose-5 phosphate is then converted to its pentose isomers, ribose-5-phosphate, and xylulose-5-phosphate. The pentose isomers ribose-6-phosphate and xylulose-5 phosphate then combine under the enzyme transketolase to make the septose sedoheptulose-7-phosphate and the triose glyceraldehyde-3-phosphate. These then combine back under the enzyme transaldolase to make the tetrose erythrose-4-phosphate and fructose-6-phosphate, completing the pentose phoshate shunt pathway. This is a strange pathway, taking a 6-carbon sugar structure and cutting it down to a 5-carbon sugar and its isomers. It takes two 5-carbon isomers ($2 \times 5 = 10$) and combines them and makes a 7-carbon + 3 carbon sugar structure ($7 + 3 = 10$). It then combines these two sugars to make a 6-carbon + 4 carbon sugar structure ($6 + 4 = 10$). Making all these sugar combinations with 3, 4, 5, 6, and 7 carbons does serve a synthetic purpose.

Moving on from sugar metabolism, we now discuss fatty acid metabolism. Yes, fatty acids and fatty acid complexes or lipids do make people fat. They are, however, an important form of energy storage in humans. Multiple fatty acid joint together in triglycerides, an ester of glycerol and three fatty acids, and two fatty acids link up in phospholipids (Figure 1.1.2). Just as glycogen stores glucose in muscles and the liver for when an energy source is needed, fats do the same all over the body. Indeed, you have fat or obese people who have stored much too much energy in the form of fat all over the body.

Let us start by introducing three molecules that assist enzymes: coenzyme A (called CoA), acyl carrier protein (called ACP), and carnitine. CoA is a coenzyme used in the synthesis and oxidation of fatty acids, and the oxidation of pyruvate in the citric acid cycle. Coenzyme A is, in chemical terms, a thiol, which can react with carboxylic acids to form thioesters, thus functioning as an acyl group carrier. It assists

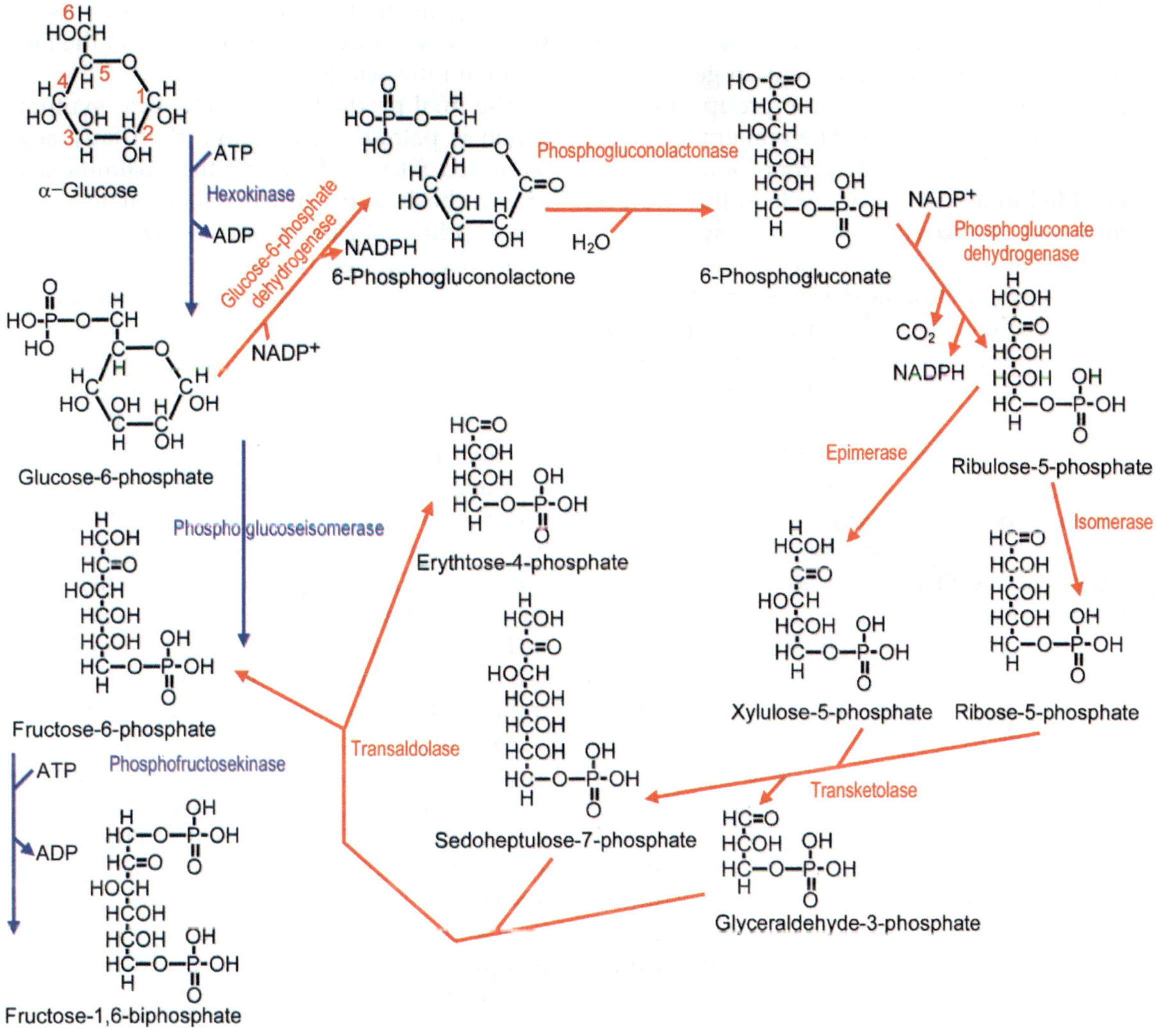

FIGURE 1.3.5 The pentose phosphate (red lines) alternative to glycolysis (blue lines).

in transferring fatty acids from the cytoplasm to mitochondria. The ester of coenzyme A that carries an acetyl group is referred to as acetyl-CoA. Mitochondrial and cytoplasmic acetyl-CoA- are seemingly the end-products of the glycolysis pathway, the fatty acid synthesis pathway, and the fatty acid oxidation pathway, and are the substrate of the citric acid cycle. ACP is an important carrier protein in fatty acid metabolism. Carnitine is a quaternary ammonium compound biosynthesized from the amino acids lysine and methionine, an important carrier for transfer of fatty acids. You need to be familiar with each of these carrier proteins, **ACP**, **carnitine**, and **CoA**.

Sugars are transferred into fats for storage through the mitochondrial citric acid pathway. Sugars are transferred by glycolysis into the

3-carbon molecule pyruvic acid. Pyruvic acid, a 3-carbon triose, enters the mitochondria where it is converted by pyruvate dehydrogenase into acetyl-CoA, a 2-carbon acidic group linked to the carrier coenzyme A. The remaining carbon is lost as CO_2. The acetyl group of acetyl CoA is added in the citric acid cycle to the 4-carbon molecule oxaloacetate in citrate synthase to make citric acid, a 6-carbon acid. If fatty synthesis is needed, citric acid leaves the mitochondrion. The enzyme ATP citrate lyase promotes the breakdown of citric acid in the cytoplasm (Figure 1.3.6), releasing oxaloacetate and acetyl CoA at the expense of an ATP molecule. Cytoplasmic acetyl CoA is the substrate for fatty acid synthesis (Figure 1.3.6). As such, if somebody drinks a Coca-Cola, the sugars may be converted to fats through pyruvic acid and mitochondrial citric acid and then converted back to cytoplasmic acetyl-CoA.

Fatty acid synthesis is somewhat of a complicated cytoplasmic pathway (Figure 1.3.6). Acetyl CoA, the starting point with two keto acid carbons, is first converted to malonyl CoA with three keto acid carbons. The two keto acid CoA esters, acetyl CoA and malonyl CoA, then adopt a different carrier molecule, ACP by the use of an *S*-malonyltransferase and *S*-acetyltransferase (Figure 1.3.6). The resulting malonyl (3-carbon)-ACP and acetyl (2-carbon) ACP then are combined using the fatty acid synthetase complex to generate a 4-carbon molecule acetoacetyl-ACP and CO_2. Three-step processing then occurs under the control of fatty acid synthetate to convert the keto group (CO) inherited from malonyl-ACP into a hydrocarbon group (HCH) leading to the generation of butyryl-ACP, a hydrocarbon fatty acid linked to ACP (Figure 1.3.6). This process is then repeated six times, each time a malonyl CoA molecule is added, each time CO_2 is released, and each time the product is processed into a 2-carbon longer fatty acid inked to ACP (Figure 1.3.6). After six repeated cycles, palmitoyl (16-carbon fatty acid)-ACP, a typical fatty acid is generated (Figure 1.3.6). This is treated with a thioesterase to remove ACP and the fatty acid palmitic acid is released.

The final product of this fatty acid synthase action is palmitic acid, a 16-carbon fatty acid. Saturated fatty acids longer than palmitic acid are synthesized by an elongation enzyme system. Additional carbons are added in 2-carbon units (CO_2 released) using malonyl CoA as the carbon donor and different elongation enzymes. In plants and animals, fatty acids can be unsaturated (get double bonds). Desaturation of fatty acids involves a process that requires molecular oxygen and NADH. The reaction results in the oxidation of both the fatty acid and NADH. The most common desaturation reactions involve the placement of a double bond between carbons 9 and 10, as occurs in the conversion of palmitic acid to palmitoleic acid, facilitated by the action of Δ9-desaturase.

The desaturation pathway utilizes desaturases to synthesize unsaturated fatty acids from full-length saturated fatty acid substrates.

Fatty acids are degraded by ß-oxidation in the mitochondrion. This is a high energy-generating pathway, making lots of NADH and $FADH_2$ used for oxidative phosphorylation (Figure 1.3.7). The ß-oxidation of palmitic acid can generate 8 NADH and 8 $FADH_2$ molecules, which can lead to the generation of 40 ATP molecules by oxidative phosphorylation before the citric acid cycle. Both fatty acid synthesis and fatty acid ß-oxidation are complicated by the extreme carrier protein needs.

In fatty acid mitochondrial ß-oxidation, the fatty acid has to first cross the double membrane barrier of the mitochondrial matrix (Figure 1.3.7). Palmitic acid is first made of a fatty acyl-CoA ester, palmitoyl-CoA. This can cross the outer membrane barrier of the mitochondrion. The fatty acyl-CoA then is converted by the enzyme carnitine acyltransferase to fatty acyl-carnitine. This is needed in order to cross the second membrane and to enter the mitochondrial matrix. Inside the matrix, the carnitine group is removed as replaced with the original CoA group (Figure 1.3.7).

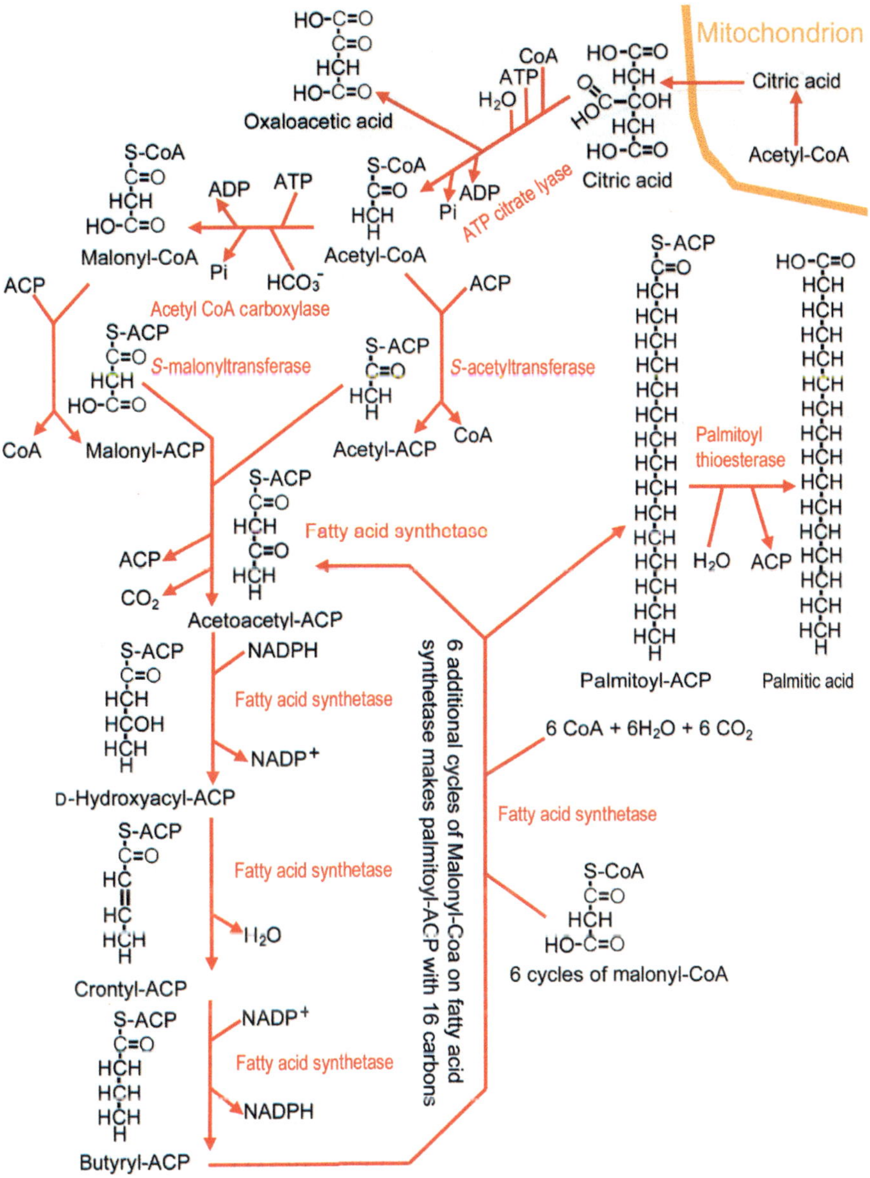

FIGURE 1.3.6 Fatty acid synthesis. ACP is acyl carrier protein, CoA is coenzyme A.

ß-oxidation like fatty acid synthesis is a 2-carbon cyclic pathway. Two carbons at a time of the fatty acids are enzymatically converted to an acetate-like keto acid CoA ester (CH_2CO) through a four-step process. Then the 2 processed carbons are released as acetyl-CoA and the entire process with fatty acyl-CoA (minus 2 carbons) repeats itself. This ß-oxidation process fills the mitochondrion with acetyl-CoA, the substrate for the mitochondrial citric acid cycle.

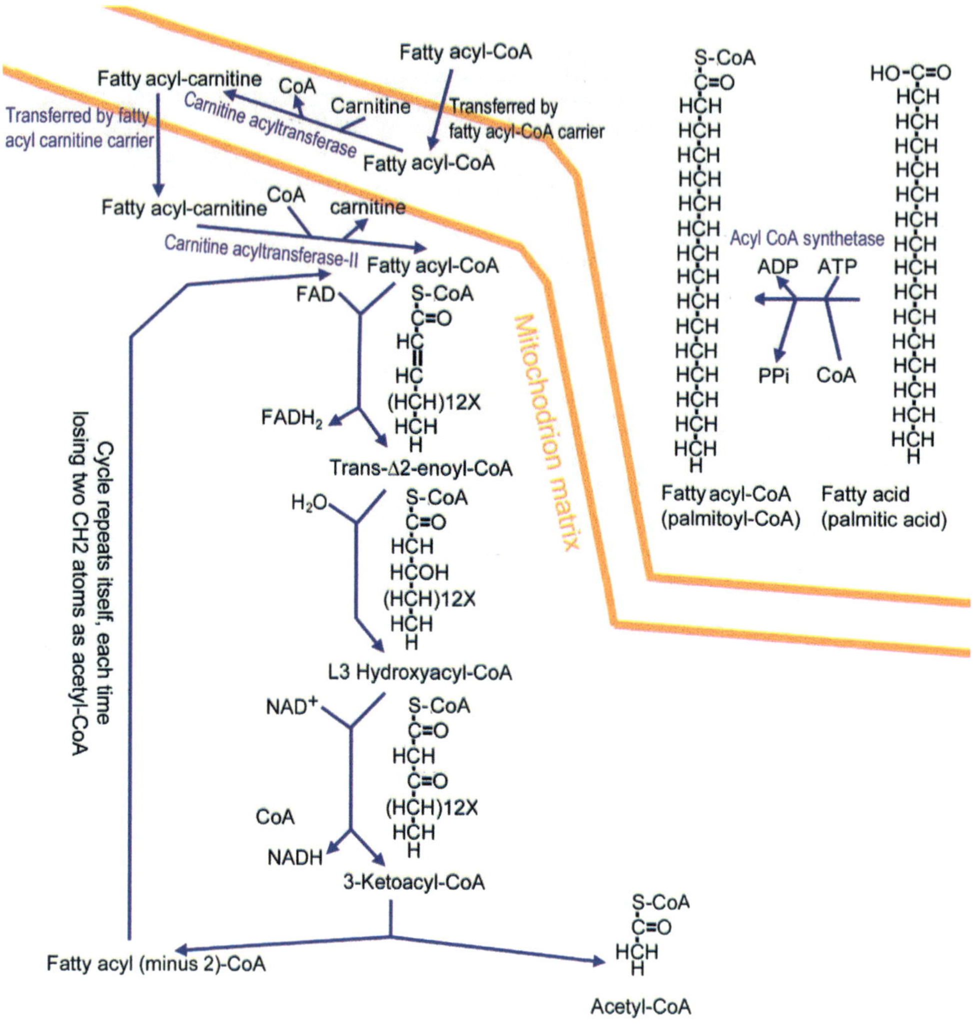

FIGURE 1.3.7 Fatty acid ß-oxidation to acetyl-CoA in the mitochondrion.

The citric acid cycle is a pathway designed to burn away carboxylic acids as two moles of CO_2. Acetyl CoA, the product of fatty acid ß-oxidation, is burned away as CO_2, as is pyruvic acid the product of sugar metabolism. Each round of the citric acid cycle generates 3 NADH, 1 $FADH_2$, and 1 molecule of ATP adding greatly to body energetics. Looking at fatty acids, one molecule of palmitic acid through ß-oxidation and oxidative phosphorylation becomes 40 ATP from ß-oxidation (8 NADH + 8 $FADH_2$). It becomes 96 ATP from the resulting 8 rounds of citric acid cycles (24 NADH, 8 $FADH_2$, and 8 ATP), or after oxidative phosphorylation, or 136 ATP total. Looking at sugars, one molecule of glucose generates 4 NADH while being converted to acetyl-CoA plus a balance of 2 ADP. This generates 14 ATP total through oxidative phosphorylation. Through two rounds of the citric acid cycle this generates 6 NADH, 2

$FADH_2$, and 2 ATP total. After oxidative phosphorylation, this is 24 ATP total. As such glucose generates 38 ATP total, a fraction of the ATP is generated from fatty acids.

Pyruvic acid is first dehydrogenated to acetyl-CoA in the mitochondrion losing a carbon as CO_2. This is the action of pyruvate dehydrogenase. Acetyl-CoA is then continuously pumped into the citric acid cycle (Figure 1.3.8). Basically, the 4-carbon keto acid oxaloacetic acid attaches to acetyl CoA to make the 6-carbon citric acid. The 6-carbon molecule loses 1 carbon as CO_2 becoming the 5-carbon keto acid α-ketoglutarate. A second carbon is then lost as CO_2 making the 4-carbon structure succinyl-CoA. The remaining steps taking succinyl-CoA back to oxaloacetic acid involve resetting this basic keto acid (Figure 1.3.8). The citric acid cycle generates GTP, which is readily converted by its nucleoside-diphosphate kinase into ATP. The mitochondrial citric acid cycle is an amazing pathway efficiently cycling acetyl-CoA and very competently generating NADH for oxidative phosphorylation. The citric acid cycle is probably the central pathway to mitochondrial metabolism.

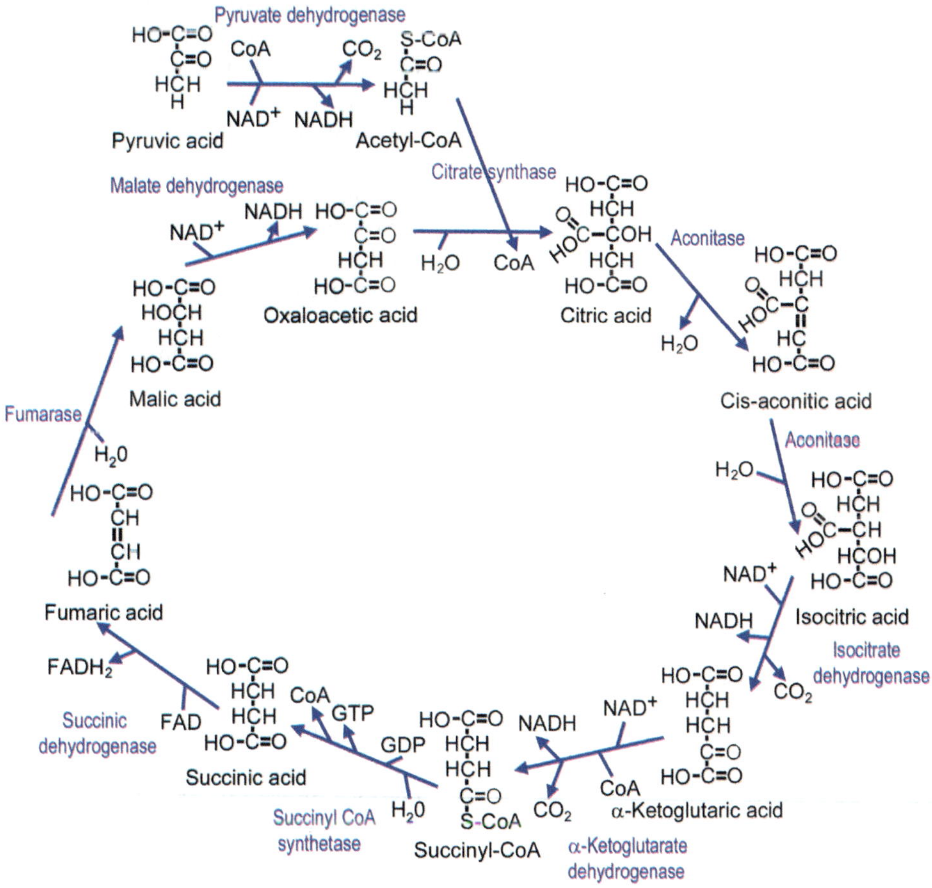

FIGURE 1.3.8 The citric acid cycle in mitochondrion.

CHAPTER KEYWORDS

Know these keywords and what each word refers to as well as its function. Read over the list. Are all these keywords familiar?

1,2-Bisphosphoglycerate
2-Phosphoglycerate
3-Ketoacyl-CoA
3-Phosphoglycerate
6-Phosphogluconate
6-Phosphogluconolactone
Acetoacetyl-ACP
Acetyl-CoA Carboxylase
Acetyl-ACP
Acetyl-CoA
Aconitase
ACP
Acyl carrier protein
Acyl CoA synthetase
ADP
Aldolase
ATP
ATP citrate lyase
ATP synthase
Butyryl-ACP
Carnitine
Carnitine acyltransferase
Carnitine acyltransferase-II
Cis-aconitic acid
Citrate synthase
Citric acid
Citric acid cycle
CoA
Coenzyme A
Crotonyl-ACP
D-Hydroxyacyl-ACP
Dihydroxyacetone phosphate
Enolase
Epimerase
Epinephrine
Erythrose-4-phosphate
FAD
$FADH_2$
Fatty acid
Fatty acid ß-oxidation
Fatty acid synthesis
Fatty acid synthetase
Fatty acyl-carnitine
Fatty acyl-CoA
Fructose-1,6-biphosphate
Fructose-6-phosphate
Fumarase
Fumaric acid
Glucagon
Gluconeogenesis
Glucose
Glucose-1-phosphate
Glucose-1-phosphate uridyltransferase
Glucose-6-phosphate
Glucose-6-phosphate dehydrogenase
Glyceraldehyde phosphate
Glyceraldehyde phosphate dehydrogenase
Glycogen
Glycogen phosphorylase
Glycogenesis
Glycogenolysis
Glycolysis
GTP
Hexokinase
Insulin
Isocitrate dehydrogenase
Isocitric acid
Isomerase
L3 Hydroxyacyl-CoA
Malate dehydrogenase
Malic acid
Malonyl-ACP
Malonyl-CoA
NAD+
NADH
Oxaloacetic acid
Palmitic acid
Palmitoyl thioesterase
Palmitoyl-ACP
Pentose phosphate shunt
Phosphoenolpyruvate
Phosphoenolpyruvate carboxykinase
Phosphofructokinase

Phosphoglucoisomerase
Phosphoglucomutase
Phosphogluconate dehydrogenase
Phosphogluconolactonase
Phosphoglycerate kinase
Phosphoglycerate mutase
Pyruvate dehydrogenase
Pyruvate kinase
Pyruvic acid
Ribose-5-phosphate
Ribulose-5-phosphate
S-Acetyltransferase
Sedoheptulose-7-phosphate
S-Malonyltransferase
Succinic acid
Succinic dehydrogenase
Succinyl-CoA
Succinyl-CoA synthetase
Trans Δ2-enoyl-CoA
Transaldolase
Transketolase
Triose phosphate isomerase
Uridine diphosphate-glucose
Uridine triphosphate
Xylulose-5-phosphate
α-Glucose
α-Ketoglutarate dehydrogenase
α-Ketoglutaric acid

INTERNET REFERENCES

Use these references to find out more.
en.wikipedia.org/wiki/Glycogen
en.wikipedia.org/wiki/Glycogen_phosphorylase
www.ncbi.nlm.nih.gov/pubmed/2011684
en.wikipedia.org/wiki/Sucrose
www.ncbi.nlm.nih.gov/pmc/articles/PMC292710/
en.wikipedia.org/wiki/Lactase
users.bergen.org/dondew/bio/AnP/Anp1/AnP1Tri1/CARB_ART/LACTOSE_SYNTH/LactoseSynth.html
en.wikipedia.org/wiki/Glycolysis
www.elmhurst.edu/~chm/vchembook/601glycolysissum.html
www.uic.edu/classes/bios/bios100/lecturesf04am/lect12.htm
en.wikipedia.org/wiki/Gluconeogenesis
www.ncbi.nlm.nih.gov/books/NBK21150/
en.wikipedia.org/wiki/Fatty_acid_synthesis
oregonstate.edu/dept/biochem/hhmi/hhmiclasses/biochem/lectnoteskga/2kjan19lecturenotes.html
en.wikipedia.org/wiki/Fatty_acid_metabolism
oregonstate.edu/dept/biochem/hhmi/hhmiclasses/biochem/lectnoteskga/2kjan14lecturenotes.html
en.wikipedia.org/wiki/Citric_acid_cycle
www.wiley.com/college/pratt/0471393878/instructor/animations/citric_acid_cycle/index.htm
en.wikipedia.org/wiki/Acetyl-CoA
en.wikipedia.org/wiki/Pentose_phosphate_pathway

CHAPTER QUIZ

Answer these questions. Do you know the answers, holy cow! Check yourself out in this quiz, and prove that you are a humanology wiz.

1. What biochemical pathway uses the enzyme glucose-1-phosphate uridyltransferase?
   - **a.** Glycolysis
   - **b.** Glycogenolysis
   - **c.** Gluconeogenesis
   - **d.** Pentose phosphate shunt
   - **e.** Glycogenesis
2. When sugars are transferred through acetyl CoA to fatty acid synthesis, what pathway moves acetyl CoA out of the mitochondrion and into the cytoplasm?
   - **a.** Acetyl-CoA—citric acid—citric acid transfer—acetyl-CoA
   - **b.** Acetyl-CoA—oxaloacetate—oxaloacetate transfer—acetyl-CoA

   c. Acetyl CoA—acetyl carnitine—acetyl carnitine transfer—acetyl-CoA
   d. Acetyl-CoA—acetyl-ACP—acetyl-ACP transfer—acetyl-CoA
   e. Acetyl-CoA—pyruvic acid—pyruvic acid transfer—acetyl-CoA
3. How effective are NADH and $FADH_2$ in mitochondrial oxidative phosphorylation?
   a. NADH and $FADH_2$ both generate one ATP molecule
   b. While NADH makes 2 ATP $FADH_2$ generates only 1 ATP molecule
   c. Both NADH and $FADH_2$ generate 3 ATP molecules
   d. While NADH generates 3 ATP molecules $FADH_2$ generates just 2.
   e. Neither ATP nor $FADH_2$ promotes the formation of any ATP molecules
4. The ß-oxidation of fatty acids generates what?
   a. Acetyl-CoA
   b. Pyruvic acid
   c. Malonyl-CoA
   d. Acetyl-ACP
   e. Malonyl-ACP
5. Is the citric acid cycle located in the mitochondria of cells?
   a. No
   b. Always
   c. Yes
   d. In some cells
   e. In humans and primate animals

CHAPTER

1.4

Amino Acid Metabolism

Amino acids are the miracle of the universe. It was the appearance of amino acids on earth 3.5 billion years ago that led to the start of life and the start of evolution. Amino acids have a central role in all cellular metabolism. Scientists have now discovered more than 500 amino acids in nature, but only 20 participate in DNA translation or human protein synthesis. Aside from their role in making all proteins and all enzymes, amino acids have many other biologically important functions. As discussed in this chapter, amino acids are readily metabolized to sugars and fatty acids to make NADH and ATP the energy molecules.

In 1953, Miller and Urey attempted to recreate the prehistoric Earth conditions. In a flask, they combined ammonia, hydrogen, methane, and water vapor plus electrical sparks (Miller, 1953). They found that new molecules were formed, and they identified these molecules as 11 standard amino acids. This is one possible explanation for the presence of amino acids and for the start of life, lightning acting on an atmosphere of ammonia, hydrogen, methane, and water vapor. Many other scientists have suggested that amino acids may have arrived on earth in the comets and meteorites that invaded earth at this time. This and other amino acid origin theories are all possible.

Here we consider the basic 20 amino acids that make up human beings, the 20 that are coded for by DNA codons, the 9 that are regularly made in humans, and the 11 others made in other animals and plants (Figure 1.4.1). A diet of meat and milk is a diet of protein. Proteins are broken down by the acid and digestive enzyme in the stomach and abdomen releasing the 20 amino acids. These are all absorbed into the blood. Digestive and degradative pathways are examined, most of which involve NADH and ATP and lead to the citric acid cycle, or generate ATP energy.

As shown in Figure 1.4.1, the 20 common amino acids range in molecular weight from 75 (Glycine) to 204 (Tryptophan). The 20 include two acidic amino acids having acidic side chain (Aspartic and Glutamic acid), two basic amino acids with amine side chains (Lysine and Arginine), two amino acids with hydroxyl side chains (Serine and Threonine), three amino acids that are aromatic in structure (Tyrosine, Phenylalanine, and Tryptophan), and four hydrophobic amino acids (valine, leucine, isoleucine, and methionine) that hide away from water in the center of a structure.

Human Physiology, Biochemistry and Basic Medicine
http://dx.doi.org/10.1016/B978-0-12-803699-0.00017-7

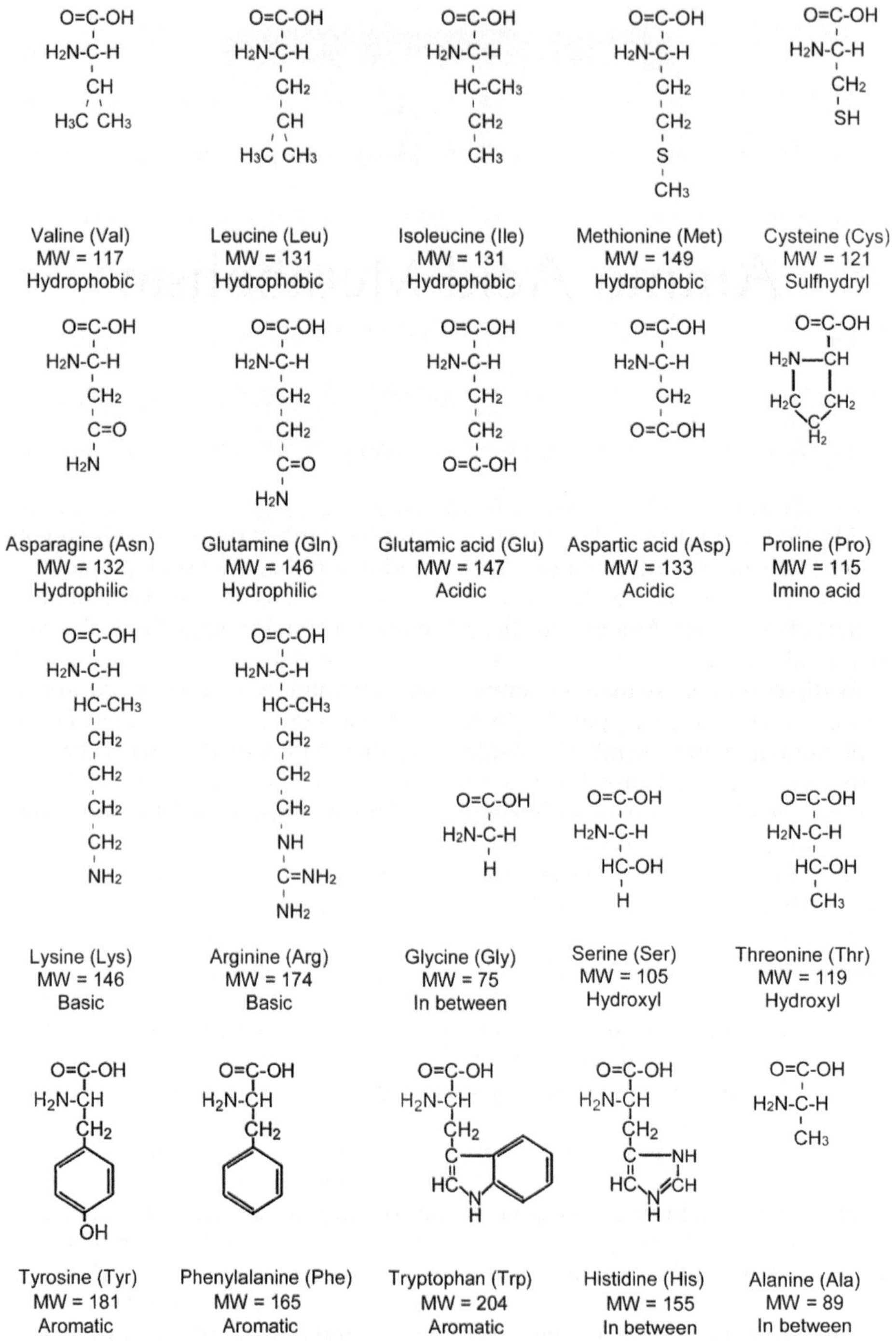

FIGURE 1.4.1 The properties of the 20 common amino acids.

Here we look carefully at the synthetic and degradative pathways of the nine amino acids that are made in humans: Ala, Asp, Asn, Cys, Glu, Gln, Gly, Pro, and Ser. We first examine the five amino acids that are synthesized by off-shoot pathways of the citric acid cycle, which are degraded into this cycle (Figure 1.4.2).

As shown in Figure 1.4.2, α-ketoglutaric acid of the citric acid cycle can be enzymatically transaminated with alanine to make glutamic acid and release pyruvic acid. The transaminase enzyme literally moves an amino group from one amino acid to a second. In an unusual reaction using ammonia, the enzyme glutamine synthetase can make glutamine from glutamate by adding an amine side chain. This is how two amino acids are freshly made in humans. To break down glutamine back to glutamic acid, the molecule is enzymatically cleaved and ammonia is released. Glutamic acid can be acted on by glutamate dehydrogenase, a separate degradation enzyme to the glutamic acid synthesis enzyme, to release ammonia and generate α-ketoglutaric acid. The ammonia generated by cells is converted to urea and secreted.

As also shown in Figure 1.4.2, oxaloacetic acid can also be transaminated to generate the amino acid aspartic acid. This transamination or passage of an amino group involves glutamic acid as the amino group. Similarly, just as glutamic acid reacts with ammonia to make glutamate, so does aspartic acid react with ammonia to

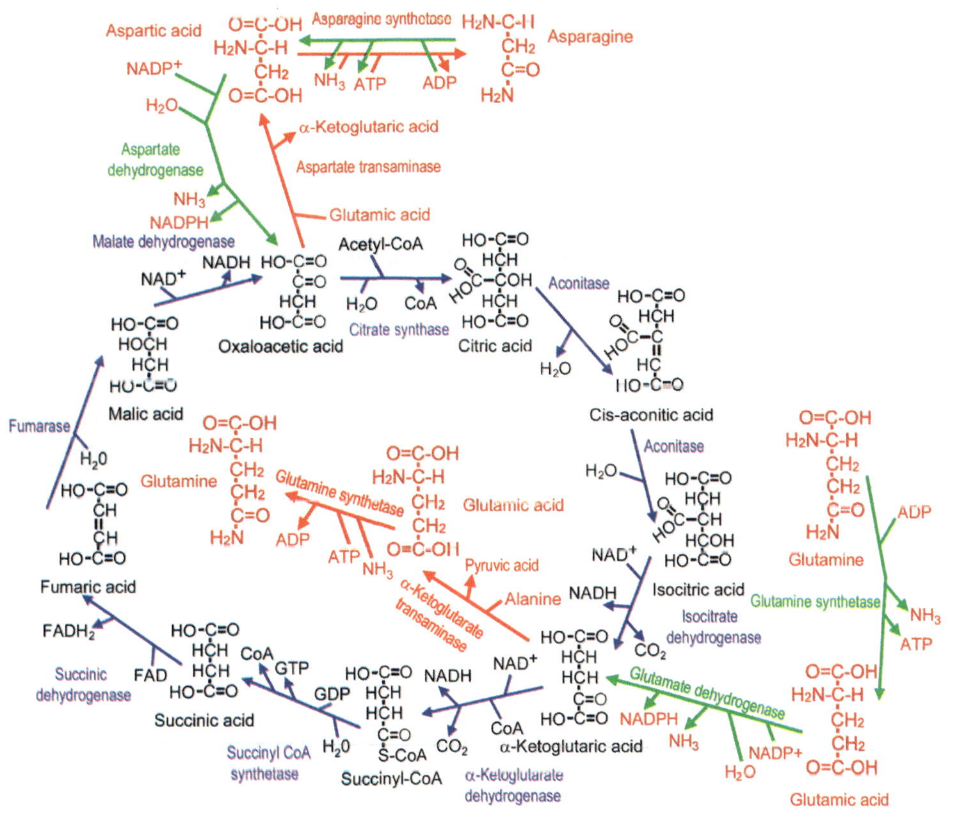

FIGURE 1.4.2 The Aspartic acid and glutamic acid channels and the citric acid cycle. Synthesis pathways are shown with red lines and degradation pathways with green lines.

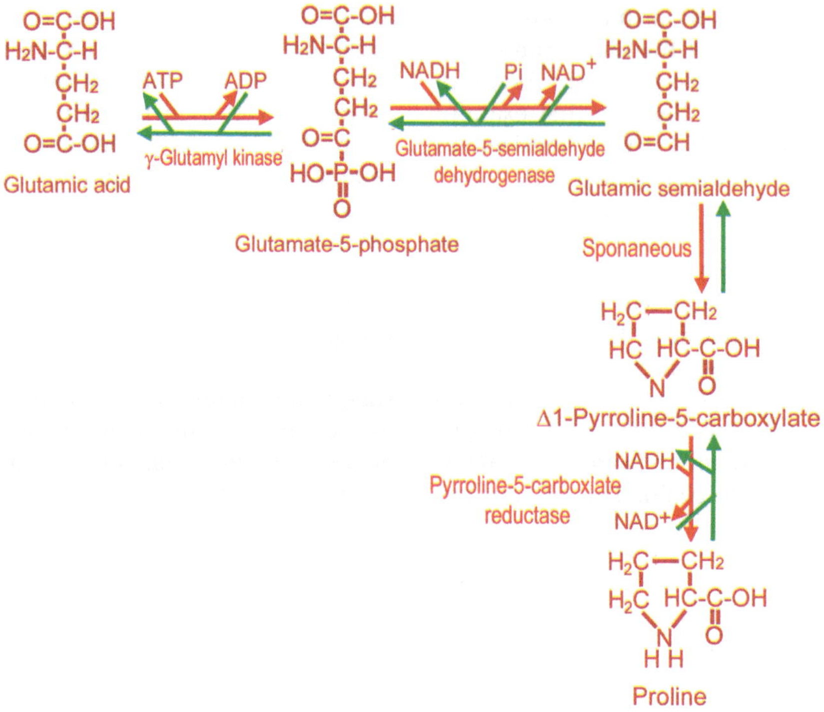

FIGURE 1.4.3 Proline synthesis (red lines) from glutamic acid and degradation (green lines) to glutamic acid.

make asparagine using the enzyme asparagine synthetase. These pathways synthesize two more amino acids. Aspartic acid like glutamic acid is degraded with the enzyme aspartate dehydrogenase, converting aspartic acid to oxaloacetic acid releasing ammonia.

The amino acid proline can be made in humans from the amino acid, glutamic acid. Glutamic acid is converted to glutamate-5-phosphate and is converted to glutamic semialdehyde, which undergoes a spontaneous folding forming the unusual pentagonal structure found in proline (Figure 1.4.3). Finally, this is acted on by a pyrroline-5-carboxylate reductase to form the amino acid proline. Proline is both synthesized by this pathway and degraded into glutamic acid and the citric acid cycle using this pathway (Figure 1.4.3).

The amino acid alanine is simply synthesized in humans and degraded through pyruvic acid, the end product of glycolysis and initiating molecule for the mitochondrial citric acid cycle. Alanine is the product of alanine transaminases on pyruvic acid (Figure 1.4.4).

The amino acids serine, cysteine, and glycine all enter carbohydrate metabolism through the middle of the glycolysis pathways at 3-phosphoglycerate, and all are synthesized in humans through 3-phosphoglycerate (Figure 1.4.5). As illustrated in Figure 1.4.5, 3-phosphoglycerate is converted

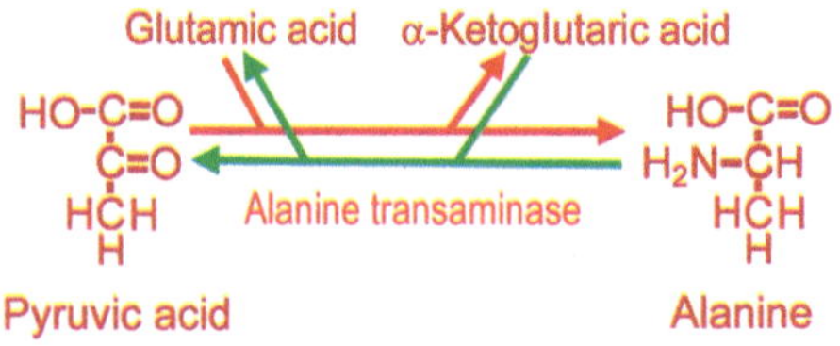

FIGURE 1.4.4 The inter-conversion of alanine and pyruvic acid.

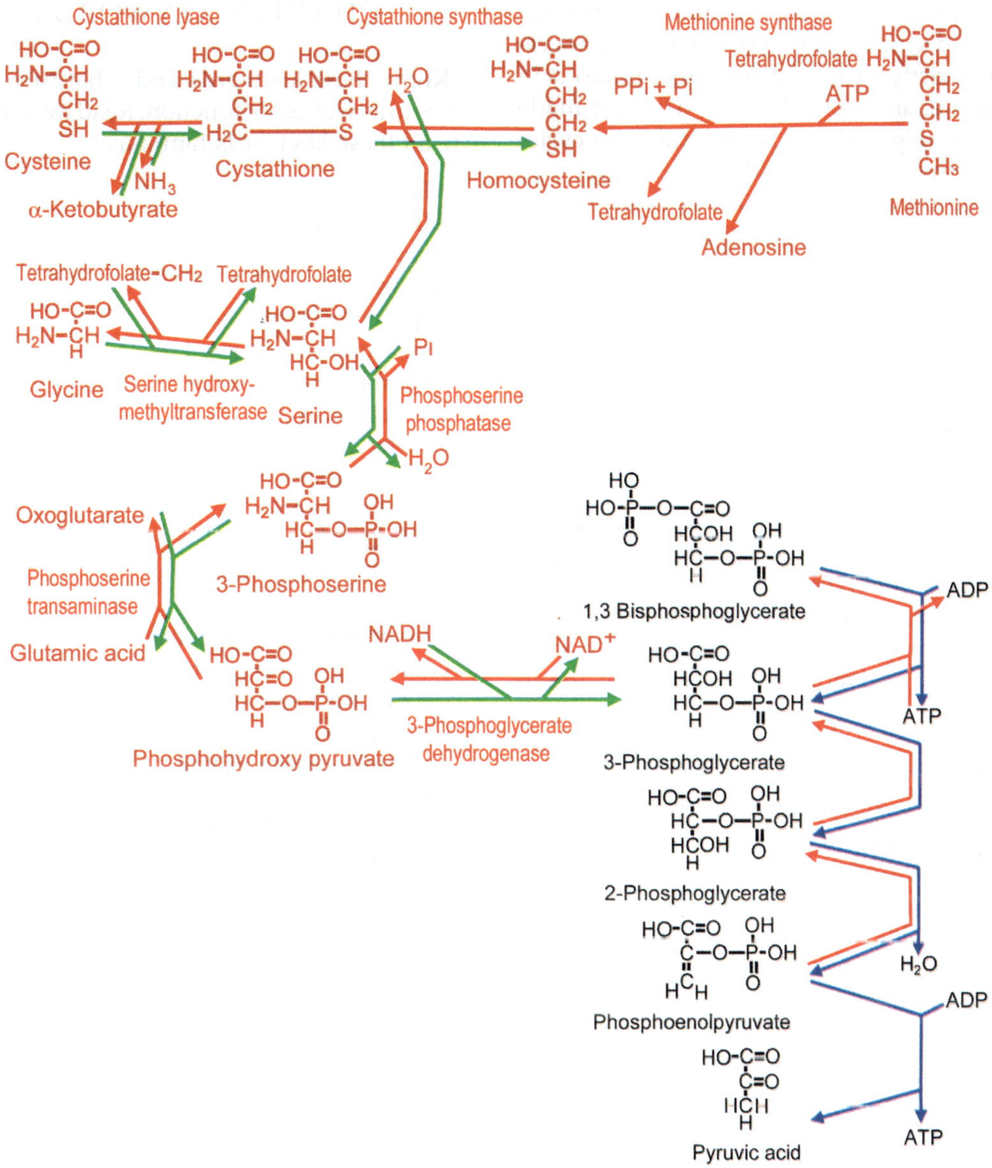

FIGURE 1.4.5 Serine, Glycine, and Cysteine synthesis (red lines) and degradation (green lines) from the glycolysis pathway.

to phosphohydroxyl pyruvate, and a specific transaminase then adds an amino group to this phospho-sugar to make the phospho-amino acid 3-phosphoserine. A final enzyme phosphoserine phosphatase hydrolyzes the phosphate group to make the amino acid serine. The amino acid serine is readily concerted to the amino acid glycine by serine hydroxymethyltransferase; this enzyme requires a vitamin derivative, tetrahydrofolate, to accept a CH_2 or methyl group.

Cysteine is also made in humans from the amino acid serine and homocysteine. Homocysteine is strictly produced from the amino acid methionine, an essential amino acid not made in humans (Figure 1.4.5). Serine and homocysteine combine to form cystathionine, a large two amino group complex, which is catalyzed by cystathionine synthase. Cystathinione is lysed to make the amino acid cysteine.

Several conclusions can be made regarding the nine amino acids made in humans. First, they are all made by side pathways of the glycolysis pathways and by side pathways of the citric acid cycle. All nine amino acids can be degraded to the glycolysis pathways and citric acid cycle and used as an energy source to generate NADH and ATP as alternatives to sugars. It is noteworthy that the synthesis of all nine amino acids involves a transaminase reaction that transfers an amino group from an established amino acid to an acidic molecule to form a new amino acid molecule; this is called amino donation.

This chapter concentrates on Glu, Gln, Pro, Asp, Asn, Ala, Ser, Gly, and Cys, the nine amino acids synthesized in humans. The remaining 11 amino acids are essential amino acids contributed by the diet, Tyr, Phe, His, Trp, Lys, Arg, Val, Leu, Ile, Met, and Thr. Meat, dairy, and other consumed products contain proteins including all 20 amino acids, essential and those synthesized. The acid of the stomach and the proteolytic enzymes produced by the stomach, pancreas, and intestines break up the proteins into separate amino acids. The proteolytic enzyme pepsin cleaves proteins at Phe, Leu, and possibly Trp, Glu, and Tyr residues. The enzyme trypsin cleaves all proteins at Lys and Arg residues, and the enzyme chymotrypsin cleaves at Tyr, Trp, and Phe residues. Multiple carboxypeptidases serially cleave amino acid starting at the COOH-terminus of the molecule. The result is separated amino acids, which are absorbed into the blood and liver.

CHAPTER KEYWORDS

Know these keywords and what each word refers to as well as its function. Read over the list. Are all these keywords familiar?

Alanine
Alanine transaminase
Arginine
Asparagine
Asparagine synthetase
Aspartate dehydrogenase
Aspartate transaminase
Aspartic acid
Carboxypeptidase
Chymotrypsin
Cystathionine
Cystathionine lyase
Cystathionine synthase
Cysteine
Glutamate dehydrogenase
Glutamate-5-phosphate
Glutamate-5-semialdehyde dehydrogenase
Glutamic acid
Glutamic semialdehyde
Glutamine
Glutamine synthetase
Glycine
Histidine
Homocysteine
Isoleucine
Leucine
Lysine
Methionine
Methionine synthase
Oxaloacetic acid
Pepsin
Phenylalanine
Phosphoglycerate dehydrogenase
Phosphohydroxy pyruvate
Phosphoserine
Phosphoserine phosphatase
Phosphoserine transaminase
Proline
Pyrroline-5-carboxylate reductase

Pyruvic acid
Serine
Serine hydroxymethyltransferase
Tetrahydrofolate
Threonine
Trypsin
Tryptophan
Tyrosine
Valine
Δ1-pyrroline-5-carboxylate
α-Ketobutyrate
α-Ketoglutarate transaminase
α-Ketoglutaric acid
γ-Glutamyl kinase

INTERNET REFERENCES

Use these references to find out more.

en.wikipedia.org/wiki/Glycine
www.biology.arizona.edu/biochemistry/problem_sets/aa/glutamate.html
en.wikipedia.org/wiki/Cysteine
http://pubchem.ncbi.nlm.nih.gov/rest/chemical/l-cysteine
www.biology.arizona.edu/biochemistry/problem_sets/aa/aspartate.html
en.wikipedia.org/wiki/Serine
jn.nutrition.org/content/136/6/1652S.full
www.jbc.org/content/98/2/465.full.pdf
en.wikipedia.org/wiki/Alanine
en.wikipedia.org/wiki/Proline
http://www.ncbi.nlm.nih.gov/pubmed/18806117
http://www.llamp.net/?q=Cho%20Proline%20metabolism
en.wikipedia.org/wiki/Asparagine
en.wikipedia.org/wiki/Glutamine
http://www.asbmb.org/asbmbtoday/asbmbtoday_article.aspx?id=18036
en.wikipedia.org/wiki/Asparticacid
en.wikipedia.org/wiki/Glutamic_acid
http://en.wikipedia.org/wiki/Transaminase
http://www.britannica.com/EBchecked/topic/602366/transaminase

CHAPTER QUIZ

Answer these questions. Do you know the answers, holy cow! Check yourself out in this quiz, and prove that you are a humanology wiz.

1. Asp and Glu are made from citric acid intermediates by what reaction?
   a. By the action of a specific dehydrogenase
   b. By the direct action of NH_3 and a specific aminase
   c. By the action of a specific transaminase
   d. By the action of a specific kinase requiring ATP
   e. By the action of NH_3, ATP, and a specific synthetase
2. Glutamine and asparagine are both synthesized from glutamic acid and aspartic acid by a specific reaction.
   a. By the action of a specific dehydrogenase
   b. By the direct action of NH_3 and a specific aminase
   c. By the action of a specific transaminase
   d. By the action of a specific kinase requiring ATP
   e. By the action of NH_3, ATP, and a specific synthetase
3. Glutamic acid and aspartic acid are both degraded, release NH_3, and enter the citric acid cycle by what action?
   a. The action of a specific dehydrogenase
   b. The direct action of NH_3 and a specific aminase

c. By the action of a specific transaminase
d. The action of a specific kinase requiring ATP
e. The action of NH_3, ATP, and a specific synthetase

4. Serine can be converted in humans by enzymatic reactions to other amino acids by
a. Gly only
b. Gly and Cys only
c. Gly, Cys, Ala, Glu, Gln, Asp, Asn, Pro only
d. Lys, Glu, Gln
e. Gly, Cys, and Ala only

5. Proline is directly enzymatically synthesized (one enzyme needed) from
a. Pyrroline-5-carboxylate
b. Tryptophan-2-acetate
c. 3-Phosphoglutamate
d. Glyceraldehyde-3-phosphate
e. 4-Phosphoaspartate

CHAPTER

1.5

DNA and Human Genetics

DNA or deoxyribonucleic acid (DNA) encodes all the genetic instructions used in the evolution and function of all known living organisms. DNA is made up of nucleotides, guanine, adenine, thymine, or cytosine nucleotides (Figure 1.5.1) attached to a sugar called deoxyribose and a phosphate group.

It is important to not confuse terms. Guanine, adenine, thymine, and cytosine are four nitrogenous nucleobases of DNA. When a base is attached to a carrier sugar like deoxyribose in DNA and ribose in RNA (ribonucleic acid), it becomes a nucleoside and takes the name deoxy-adenosine, deoxy-guanosine, deoxy-thymidine, and deoxy-cytidine in DNA or adenosine, guanosine, cytidine, and uridine (uridine always replaced thymidine in RNA) as in RNA. With a phosphate group added, the nucleoside becomes a nucleotide and becomes deoxy-adenosine monophosphate (A), deoxy-guanosine monophosphate (G), deoxy-thymidine monophosphate (T), and deoxy-cytidine monosphosphate (C) as in DNA and adenosine monophosphate (A), guanosine monophosphate (G), uridine monophosphate (U), and cytidine monosphosphate (C) as in RNA. It is strictly the nucleotide that comprises DNA and RNA (Figure 1.5.1).

The DNA and RNA nucleotides are joined to one another in a chain in DNA and RNA by covalent bonds between the sugar of one nucleotide and the phosphate of the next, resulting in an alternating sugar-phosphate backbone. In DNA only, according to base-pairing rules, adenine nucleotides always base pair with thymidine nucleotides, and guanine nucleotides always base pair with cytosine nucleotides. The base pairing is caused by hydrogen bonds between the nucleotides (Figure 1.5.1). If one linked strand of DNA comprises a chain of -G-A-T-C-G-A-T-C-, it pairs with a second strand, according to base pairing, and thus comprises the corresponding sequence -C-T-A-G-C-T-A-G-. The result is that the two strands or base pairs are attached to each other in DNA. The two strands fold around each other and form the classical DNA double-helix structure (Figure 1.5.2).

The base-paired strands of DNA effectively store the same biological information -G-A-T-C-G-A-T-C- storing the "sense" and -C-T-A-G-C-T-A-G- storing the "antisense" sequence of the genetic code. Biological information is effectively replicated as the two strands are separated. The DNA sequence contains the genetic code and is translated into amino acids.

Human Physiology, Biochemistry and Basic Medicine
http://dx.doi.org/10.1016/B978-0-12-803699-0.00018-9

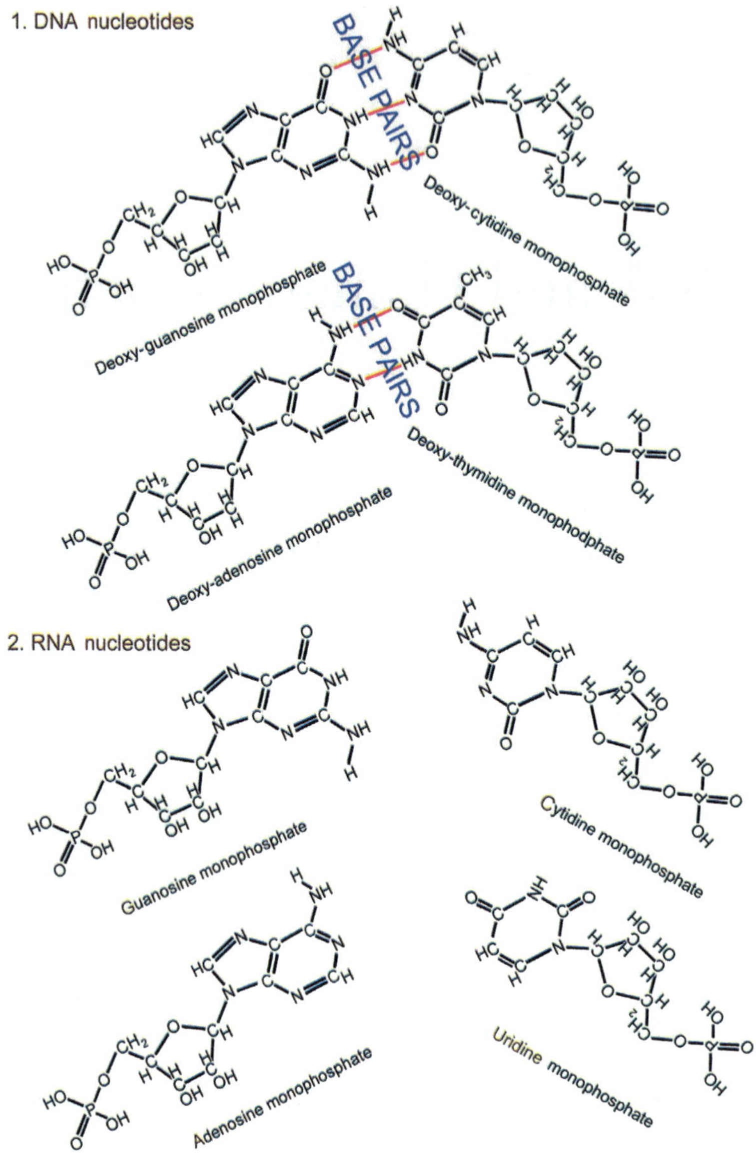

FIGURE 1.5.1 Nucleotide structure of DNA and RNA.

A significant portion of the DNA sequence, more than 98% for humans, is noncoding, i.e., these sections do not serve a function of encoding amino acids.

Within cells, DNA is arranged in long structures called chromosomes, each containing 100-220 million base pairs (A, G, T, C sequence nucleotides). During cell division, these chromosomes are duplicated, providing each cell with its own complete set of chromosomes. Humans store most of their DNA inside the cell nucleus and some of their DNA in organelles, such as

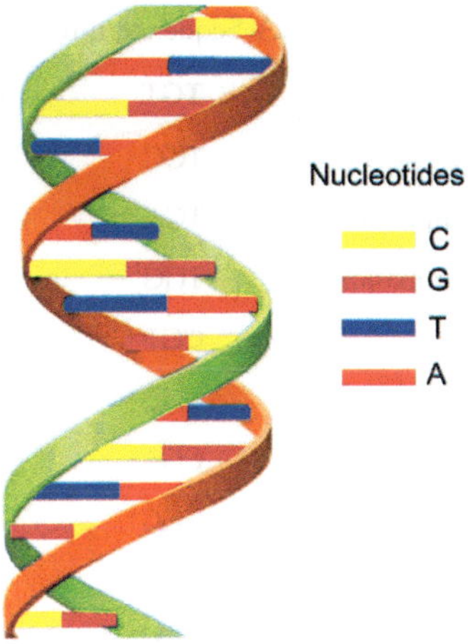

FIGURE 1.5.2 Double-stranded base-paired DNA.

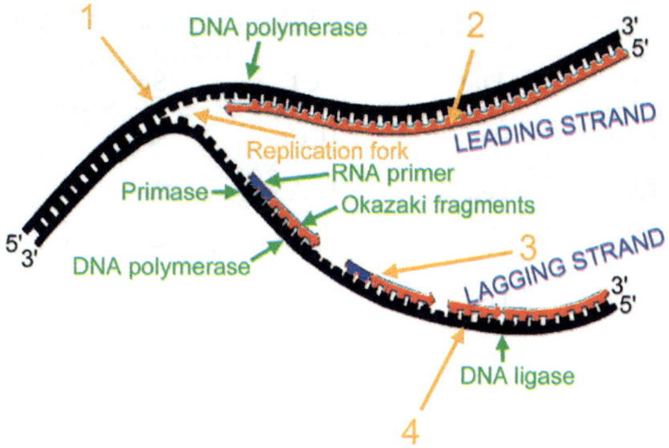

FIGURE 1.5.3 DNA replication. The black cog-like structure is double-helix DNA (left side) and the annealed sense and antisense strands (center and right side).

mitochondria. Within the chromosomes, a protein called chromatin compacts and folds around the DNA. The folded protein that surrounds and exposes DNA is called histones, and the DNA-protein complex is called nucleosomes. These compact structures guide the interactions between DNA and other proteins, controlling which parts of the DNA are transcribed and translated to make needed proteins and enzymes in a specific cell.

In the double-helix structure, the two stands of DNA run antiparallel. The asymmetric ends of DNA strands are called the 5′ (five prime) and 3′ (three prime) ends, with the 5′ end having a terminal phosphate group and the 3′ end a terminal hydroxyl group. One strand runs 3′-5′ and the base-paired strand runs in the opposite direction (Figure 1.5.3).

The DNA codes for specific amino acids in groups of three nucleotides or codons. The DNA genetic code is listed in Table 1.5.1. As shown, codon code for all 20 common amino acids. Three codons TAA, TAG, and TGA are termination codons coding for sequence finished, stop translating the sequence. A copy of the DNA sequence of a gene coding for a specific protein or enzyme is made, which is a RNA or RNA copy called messenger RNA (mRNA). The copying process is called transcription. The mRNA transcript leaves the nucleus and goes to the ribosome where it is read and translated through a transfer RNA (tRNA) into a tRNA conveyed amino acid sequence. This process is called translation.

A gene sequence contains coding sequences for making proteins and enzymes or "exons" and noncoding or intervening sequences or "introns." An intron is any nucleotide sequence within a gene that is removed by RNA splicing while the final protein-coding mRNA is being generated. Sequences that are joined together to form the final mRNA sequence after RNA splicing are exons. When proteins are to be made from an mRNA strand, splicing takes place following transcription and preceding formal translation in the ribosome. RNA differs somewhat from DNA in not having the nucleoside thymidine. In place, RNA uses the nucleoside uridine (uracil-ribose). The four distinct nucleotides comprising RNA, in which nucleosides are formed with the sugar ribose rather than deoxyribose, are adenosine monophosphase (A), guanosine monophosphate (G), cytidine monophosphase (C), and uridine (uracil-ribose) monophosphate (U). Uridine nucleotides like thymidine nucleotides base pair with adenosine nucleotides. The mRNA genetic code is

TABLE 1.5.1 The Triplet Codon DNA Genetic Code, Translating the "Sense" DNA Sequence into Amino acids

TTT	Phe	TCT	Ser	TAT	Tyr	TGT	Cys
TTC	Phe	TCC	Ser	TAC	Tyr	TGC	Cys
TTA	Leu	TCA	Ser	TAA	Termination	TGA	Termination
TTG	Leu	TCG	Ser	TAG	Termination	TGG	Trp
CTT	Leu	CCT	Pro	CAT	His	CGT	Arg
CTC	Leu	CCC	Pro	CAC	His	CGC	Arg
CTA	Leu	CCA	Pro	CAA	Gln	CGA	Arg
CTG	Leu	CCG	Pro	CAG	Gln	CGG	Arg
ATT	Ile	ACT	Thr	AAT	Asn	AGT	Ser
ATC	Ile	ACC	Thr	AAC	Asn	AGC	Ser
ATA	Ile	ACA	Thr	AAA	Lys	AGA	Arg
ATG	Met*	ACG	Thr	AAG	Lys	AGG	Arg
GTT	Val	GCT	Ala	GAT	Asp	GGT	Gly
GTC	Val	GCC	Ala	GAC	Asp	GGC	Gly
GTA	Val	GCA	Ala	GAA	Glu	GGA	Gly
GTG	Val	GCG	Ala	GAG	Glu	GGG	Gly

essentially identical to the DNA genetic code except that U replaces T. mRNA does not base pair, and double-strand like DNA and remains single-stranded.

Four distinct DNA and RNA pathways occur: "replication," in which the DNA sequence is duplicated as needed in cell division, "transcription," in which mRNA makes a copy of a gene sequence, mRNA splicing in which exons are dissected and "translation" in which the mRNA sequence is converted into the coded amino acids in ribosomes.

REPLICATION

Replication is the protocol whereby DNA or genes or chromosomes duplicate themselves when cells divide. The process of DNA replication is illustrated in Figure 1.5.3 and described by steps and numbers. In the first step, an enzyme called a helicase breaks apart the base-paired two DNA strands. The function of helicases is to unpackage the DNA genes. They are motor proteins that move directionally along the DNA double helix separating the two annealed nucleic acid strands using energy derived from ATP hydrolysis (Figure 1.5.3 Arrow 1).

The DNA separates into a leading strand of DNA, which is replicated in the same direction as the replication fork, and a lagging strand of DNA, which is replicated in the opposite direction of the replication fork. Because of its orientation, replication of the lagging strand is more complicated than that of the leading strand.

The replication fork is a structure that forms during DNA replication. The replication fork is created by helicases that cleave the hydrogen bonds holding the two DNA strands together. These two strands serve as the template that will be created as DNA polymerase matches complementary nucleotides to the separated strand templates. DNA is always synthesized in the 5′-3′ direction. Since the lagging strand templates are oriented in opposite directions of the replication fork, a major issue is how to achieve synthesis of new lagging strand DNA in a direction of synthesis opposite to the direction of the growing replication fork.

On the leading strand, a DNA polymerase "reads" the leading strand template and adds free complementary nucleotides to the nascent leading strand on a long continuous basis (Figure 1.5.3 Arrow 2). The DNA polymerase on the leading strand in humans is DNA polymerase epsilon. On the lagging strand, a DNA polymerase delta uses the free complimentary nucleotides to synthesize DNA in short separated segments. A primase reads the template DNA and initiates synthesis of a short complimentary RNA primer (Figure 1.5.3 Arrow 3). The RNA primer is replaced by DNA. The DNA polymerase delta extends the primed segments, forming Okazaki fragments or short fragments of DNA. The fragments of DNA are joined together by a DNA-linking enzyme called DNA ligase (Figure 1.5.3 Arrow 4). This way, two new double-helix strands of DNA are formed and a double-helix structure results.

TRANSCRIPTION

Transcription is the process whereby proteins and enzymes are made exactly as coded on the sense DNA sequence. A segment of DNA is unwound or hydrogen bonds are broken, exposing segment of the two DNA strands. An mRNA copy is made or complementary copy of only the antisense DNA strand, adenosine monophosphate bases matching up with deoxy-T bases, uridine (not thymidine since RNA) monophosphate bases matching up with deoxy-A bases, guanosine monophosphate bases matching with deoxy-C bases, and cytidine monophosphate with deoxy-G bases. The antisense mRNA dupe is effectively an mRNA copy of the DNA sense sequence except that uridine replaces thymidine nucleotides. The mRNA leaves the nucleus through nuclear pores and goes to ribosomes where the protein is constructed by tRNA molecules carrying specific amino acids.

An activator protein specifically binds the gene selecting it for transcription. Proteins called transcription factors unwind the segment of DNA strands and an RNA polymerase transcribes the single antisense strand of DNA to make the mRNA (Figure 1.5.4). The strand that serves as the template for transcription is always the antisense strand. The strand that is not

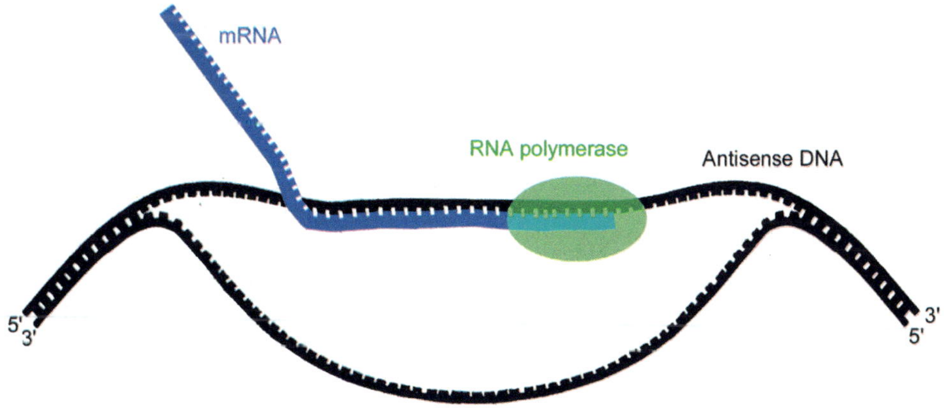

FIGURE 1.5.4 DNA transcription.

transcribed is called the sense strand. RNA polymerase moves along the DNA copying the base sequence until it reaches a terminator sequence (see Table 1.5.1). At that point, RNA polymerase releases the mRNA polymer and it detaches from the DNA (Figure 1.5.4). The 3′ terminus of the new mRNA gets a long poly-A sequence tail. As many as 100 adenosine-monophosphate residues are lined up at this terminus to identify the mRNA.

mRNA SPLICING

While the mRNA molecule copies the gene on the chromosome in the nucleus, regions to be translated as amino acids in a protein (exons) need to be separated from unread regions (introns). Splicing is done in a series of reactions that are catalyzed by the spliceosome, a complex of RNA and protein. As part of the RNA processing pathway, introns are removed by RNA splicing either shortly after or concurrent with transcription.

Within the sequences to be removed are consensus sequences that identify the introns. Introns invariably start at 5′ terminus with a GU identifier, and invariably end at the 3′ terminus with an AG identifier. Sequences identifying the introns include M-A-G-[break]-G-U-R-A-G-U, C-U-R-[A]-Y, and Y-rich-N-C-A-G. These sequences fold and tell the spliceosome exactly where to cut and rejoin the mRNA (Figure 1.5.5). The final mRNA contains only exons or only the sequence to be translated in the ribosome.

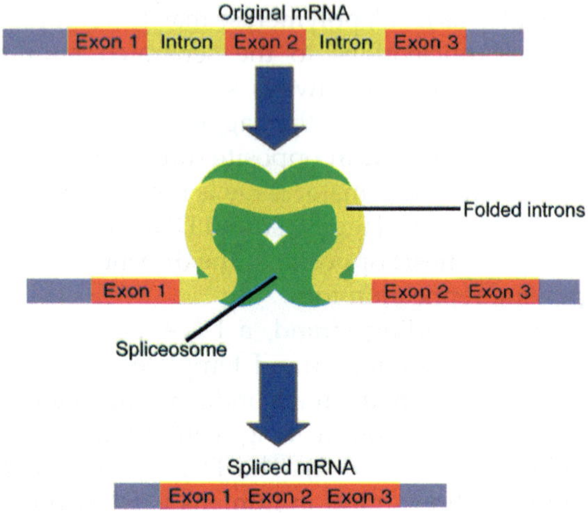

FIGURE 1.5.5 Slicesome splicing of mRNA.

TRANSLATION OF mRNA

The mRNA leaves the nucleus through nuclear pores and travels to a ribosome. Here translation occurs or conversion of RNA to an amino acid sequence occurs. The process of translation is illustrated in Figure 1.1.5 in Chapter 1 of this book.

tRNA is an adaptor molecule composed of RNA, 73-94 ribonucleotides in length, which is the link between the nucleotide sequence of nucleic acids and the amino acid sequence of proteins. This link is achieved by tRNA carrying an amino acid and being coded by a three-nucleotide antisense sequence. When mRNA codes for the amino acid Cys, for instance, the mRNA coding sequence is UGC (Table 1.5.1). A tRNA carries the amino acid Cys. It contains the antisense codon, ACG. This ACG antisense sequence binds the coding mRNA, placing the amino acid Cys in the protein (Figure 1.1.5).

The covalent attachment of the specific amino acid to a tRNA 3′ end is catalyzed by one of a group of enzymes called aminoacyl-tRNA synthetases. During protein synthesis, tRNAs with attached amino acids are delivered to the ribosome by proteins called elongation factors, eEF-1, which aid in decoding the mRNA codon sequence. If the tRNA's anticodon matches the mRNA codon, another tRNA already bound to the ribosome transfers the growing polypeptide chain from its 3′ end to the amino acid attached to the 3′ end of the newly delivered tRNA, a reaction catalyzed in ribosomes (Figure 1.1.5).

GENES AND CHROMOSOMES

Human cells contain 23 pairs of chromosomes, 23 from the individual's father and 23 from the mother. Sexually, the 23 pairs include an X and a Y chromosome in males and two X chromosomes in females. Genes code for proteins, enzymes, glycoproteins (proteins made of amino acids and then receive carbohydrate side chain in the rough endoplasmic reticulum and Golgi apparatus of cells), structural proteins, hormones and muscular proteins, and general cellular proteins.

The 23 pairs of chromosomes each contain the coding for hundreds of genes linked head to tail. Approximately 25,000 genes fill the human 23 chromosomes in total. The DNA double helix is packaged by proteins into a condensed structure called chromatin. This allows the very long DNA molecules to fit into the cell nucleus. Chromosomes must be replicated, divided, and passed successfully to their daughter cells to ensure the genetic diversity and survival of their progeny. Humans possess multiple large linear chromosomes contained in the cell's nucleus. Each chromosome has one centromere, with one or two arms projecting from it.

CHAPTER KEYWORDS

Know these keywords and what each word refers to as well as its function. Read over the list. Are all these keywords familiar?

Adenine
Aminoacyl-tRNA synthetase
Antisense DNA
Codon
Cytosine
Deoxy-adenosine
Deoxy-adenosine monophosphate
Deoxy-cytidine
Deoxy-cytidine monosphosphate
Deoxy-guanosine
Deoxy-guanosine monophosphate
Deoxy-thymidine
Deoxy-thymidine monophosphate
DNA
DNA ligase
DNA polymerase
DNA polymerase delta
DNA polymerase epsilon
Exon
Guanine
Helicase
Intro
Lagging strand
Leading strand
Messenger RNA
Nucleobase
Nucleoside
Nucleotide
Okazaki fragments
Primase
Replication fork
RNA
RNA polymerase
Sense DNA
Spliceosome
Thymidine
Transcription
Transfer RNA
Translation replication
tRNA
Uracil
Uridine

INTERNET REFERENCES

Use these references to find out more.
en.wikipedia.org/wiki/DNA
en.wikipedia.org/wiki/DNA_replication

en.wikipedia.org/wiki/Transcription_%28genetics%29
biology.about.com/od/cellularprocesses/ss/Dna-Transcription.htm
en.wikipedia.org/wiki/Transfer_RNA
www.nature.com/scitable/topicpage/translation-dna-to-mrna-to-protein-393
en.wikipedia.org/wiki/Messenger_RNA

CHAPTER QUIZ

Answer these questions. Do you know the answers, holy cow! Check yourself out in this quiz, and prove that you are a humanology wiz.

1. What is the difference between a DNA nucleoside a DNA nucleotide?
   a. DNA nucleosides are deoxy-ribose
   b. DNA nucleotides include deoxy-uridine monophosphate
   c. DNA nucleotides contain a phosphate
   d. No difference at all
   e. They are linked deoxy-ribose to phosphate group
2. In replication, which DNA strand is duplicated forming multiple Okazaki fragments?
   a. Lagging strand
   b. Leading stand
   c. 5′-3′ strand
   d. 3′-5′ strand
   e. Both strands
3. In transcription what molecule initializes the event?
   a. A helicase
   b. An activator
   c. RNA polymerase
   d. tRNA
   e. A template protein
4. What function does aminoacyl-tRNA synthetase have?
   a. Links the specific tRNA to the mRNA coding sequence
   b. Links the specific amino acid to the specific tRNA
   c. Catalyzes the synthesis of the amino acid-amino acid linkage
   d. Links the multiple aminoacyl-tRNA molecules, 3′-5′
   e. Catalyzes the translation process, rRNA-tRNA protein
5. In RNA splicing what is the exact role of the spliceosome?
   a. Carrying out the slicing process
   b. Identify the mRNA sequences to be spliced
   c. Cleaving the intron-exon linkages
   d. Degrading the introns
   e. Binding the exons

CHAPTER

1.6

Protein Structure and Folding

Chapter 1.4 covered amino acid synthesis and Chapter 1.5 covered DNA transcription and translation, both chapters leading to the formation of amino acids and strings of linked amino acids. Strings of amino acids, 3-500 residues long, form proteins. All strings of amino acids fold and form specific three-dimensional shapes, whether composed only of amino acids, glycoproteins, proteoglycans, or nucleopeptides.

Enzymes fold forming three-dimensional structures that specifically bind the enzyme substrate, then move the substrate and combine it with other bound molecules such that a product is optimally formed. Receptor proteins fold into a shape such that a hormone is specifically bound. The receptor protein is folded such that a second hormone may be released, or a messenger molecule is processed or formed. A protein hormone is folded so that it is a specific shape or messenger, and will bind a specific receptor. Proteins such as hemoglobin subunits are formed of such a shape so that they specifically bind heme molecules and oxygen and so on.

Folding occurs from the interaction between 20 different amino acids. The final structure is determined from the amino acid sequence and by any folding dictated by the ribosomal mRNA coding sequence. A protein starts folding on the ribosome and may start folding before translation of a string of amino acids is complete. The folding process varies according to the concentration of salts, the presence of cofactors, or of a molecular chaperone molecule. Important driving forces in folding are amino acid-amino acid hydrogen bonds, amino acid-amino acid hydrophobic interactions, amino acid-amino acid charge interactions, Cys-Cys disulfide bridges, and attachment of sugar side chains and other molecular modification.

Strings of amino acids form α-helix structures and β-pleated sheets, unfolded structures, twist and turns. An example of DNA sequence, corresponding amino acid sequence, and complex folded multisubunit three-dimensional structure is shown in Figure 1.6.1.

Protein three-dimensional structure can be examined by nuclear magnetic resonance spectrometry and by circular dichroism spectrometry and by crystal formation and X-ray crystallography.

Protein folding can be physically disrupted, creating inactive structures by unnatural temperatures, by extremities of pH, by mechanical forces, and by denaturants. However, under certain circumstances, proteins will naturally refold.

Aggregated misfolded proteins occur in Creutzfeldt-Jakob disease, mad-cow disease,

Human Physiology, Biochemistry and Basic Medicine
http://dx.doi.org/10.1016/B978-0-12-803699-0.00020-7

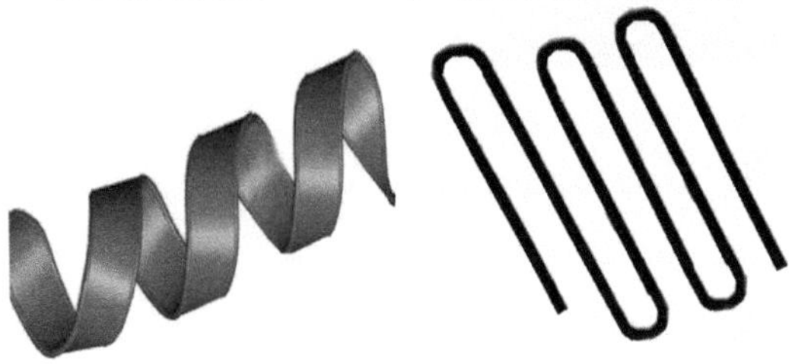

FIGURE 1.6.1 α-Helix and β-pleated sheet.

Alzheimer's disease, familial amyloid cardiomyopathy, Huntington's disease, and Parkinson's disease, but it is not clear if the misfolded proteins are the cause of the disease or a consequence. Physical protein replacement therapy corrects these disorders and is a treatment of choice.

Protein structure can be divided into four distinct levels: Primary structure is the linear amino acid sequence; secondary structure is the molecule considering α-helixes and β-pleated sheets, hydrogen bonds, and effects of Cys-Cys disulfide bridges; tertiary structure reflects the complete three-dimensional structure of the protein subunits; and quaternary structure is the complete three-dimensional structure after combining the protein's multiple subunits.

CHAPTER KEYWORDS

Know these keywords and what each word refers to as well as its function. Read over the list. Are all these keywords familiar?

Circular dichroism
Primary structure
Quaternary structure
Secondary structure
β-pleated sheet
Tertiary structure
X-ray crystallography
α-Helix

INTERNET REFERENCES

Use these references to find out more.
en.wikipedia.org/wiki/Protein_folding
en.wikipedia.org/wiki/Protein_structure
en.wikipedia.org/wiki/Protein_design
en.wikipedia.org/wiki/Protein_dynamics

CHAPTER QUIZ

Answer these questions. Do you know the answers, holy cow! Check yourself out in this quiz, and prove that you are a humanology wiz.

1. Common amino acid folding structures include:
   - **a.** γ-Double folds and β-helixes
   - **b.** β-Helixes and ε-fold backs
   - **c.** g-Double folds and α-pleated sheets
   - **d.** α-Helixes and β-pleated sheets
   - **e.** α-Folds and β-twists
2. Folding first starts
   - **a.** In the ribosome before translation is complete
   - **b.** In the nucleus before transcription is compete
   - **c.** In the golgi apparatus before glycosylation is complete
   - **d.** In the lysosome before synthesis is complete
   - **e.** On the endoplasmic reticulum before glycosylation is complete

Section 2
Human Frames

CHAPTER

2.1

The Skeleton

It is important to remember the names of the major bones in the body (Figure 2.1.1). They define the body's shape, rigidity, and finesse, and are what remains of humans after they are buried. There are many ways to remember those bone names. Here is a short riddle to help you.

Remember Those Bones

Open your mouth, on the bottom is your **mandible**, the bone of jaw.

On the top is your **maxilla**, two bones with which you suck a straw.

Don't forget that **cranium** comfortably houses your inquisitive brain.

Then there is your shoulder bones, **clavicle** and **scapula**, with which you play a football game.

There is the **sternum** at the meeting point of your **ribs**.

Then the bones of your upper arm, the **humerus**, making its comedy dibs.

Your lower arm comprises the **ulna** and the **radius**.

And your hand, **carpals** and **metacarpals**, where your **phalanges** or fingers are quite contagious.

Never forget those wide bones of your **pelvis**.

And that the giant thigh **femur** which plugs right in, like records and the many songs of Elvis.

Tibia and **fibula** are the bones of your lower legs, and the **patella** or knee bone, often gets blows.

Then **calcaneus**, **tarsals**, **metatarsals** and **phalanges** that together form your feet and toes.

It is important to know the structure of the spine or the vertebra. At the very top of the spine linked to the cranium is the first vertebra called the atlas (Cervical 1), and the second vertebra called the axis (Cervical 2). These vertebra allow the head to move or rotate. These are the first two of seven vertebra, cervical vertebra, which form the human neck. The cervical vertebra are followed by 12 thoracic vertebra. These thoracic vertebra accommodate the chest cavity and ribs. These are followed by 5 lumbar vertebra between the rib cage and the pelvis. The nerves that emanate from the spine at this point control the pelvic and legs. Finally, the spine is terminated in the pelvis by 5 fused-together sacral vertebra and a cork or useless blockage of 5 fused vertebra called the coccyx or tailbone.

The hands and feet are each made up from many bones (Figure 2.1.2). The Ulna and the rotating bone the Radius support the hands and wrists. The wrists are made up from multiple stubby bones called the Carpals. The carpals include the Scaphoid and Lunate and Triquetral bones the first line of bones attached to the Ulna and Radius, then the Trapezium, Trapezoid, Capitate and Humate bones linking the hands and fingers. The hands are composed of

http://dx.doi.org/10.1016/B978-0-12-803699-0.00002-5

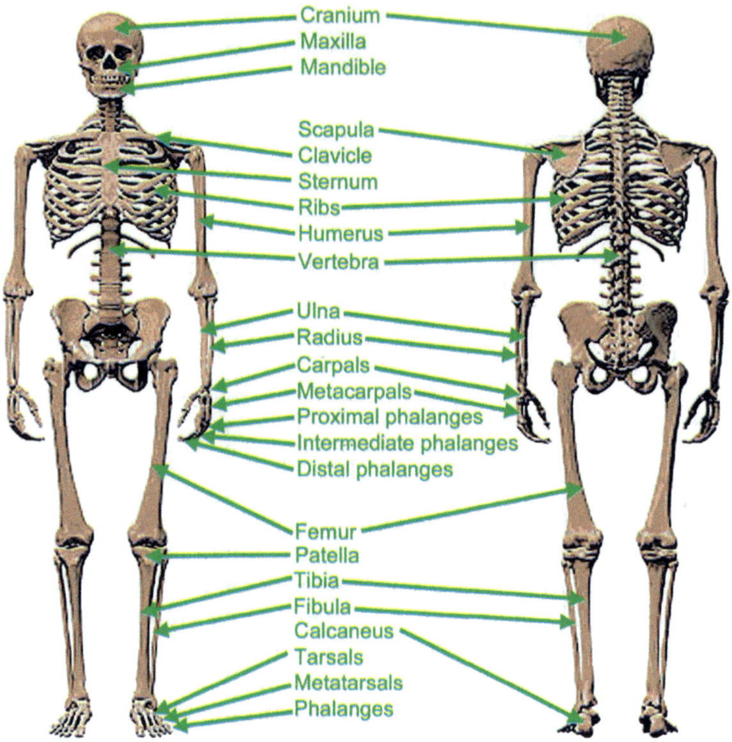

FIGURE 2.1.1 The bones.

Metacarpals, and the fingers are Proximal phalanges extending out to Middle Phalanges, to Distal phalanges (Figure 2.1.2).

The feet are very similar in structure to the hands. The bones of the feet are seven stubby bones called the Tarsals (Figure 2.1.2). These comprise two big bones at the rear of the foot, the Talus and Calcaneus, then the Navicular. A row of bones attaches the toes. These are the Cuboid, Lateral cuneiform, Intermediate cuneiform, and Medial cuneiform bones (Figure 2.1.2). The feet extend into the toes with the Metatarsals. The toes comprise, exactly like the fingers, the Proximal phalanges, Middle phalanges, and finally the Distal phalanges (Figure 2.1.2).

The thigh bone is the Femur, by far the biggest bone in the body (Figure 2.1.1). The knee joint has a unique bone called the Patella. The lower leg bones are the Tibia and the thinner Fibula adjacent to the Tibia. The Tibia and Fibula attach the leg to the Tarsals or foot bones.

The upper arm bone is the Humerus (Figure 2.1.1). The forearm bones across the elbow joint are the Ulna and the Radius. The Radius rotates as one rotates the arm and hand. The Ulna and Radius are linked to Carpals at the wrist. The shoulders comprise the clavicle or collarbone. This is a flat bone that serves as a support between the Scapula and the Sternum. It is the only long bone in the body that lies horizontally. The Scapula is the rear shoulder blade bone. The Sternum is that front center bone that links up all the ribs.

The 12 left and 12 right Ribs are the long curved bones that form the chest cage. The first seven sets of ribs, known as "true ribs," are directly attached to vertebra and to the Sternum, linked through costal cartilage. Rib 1 is a short,

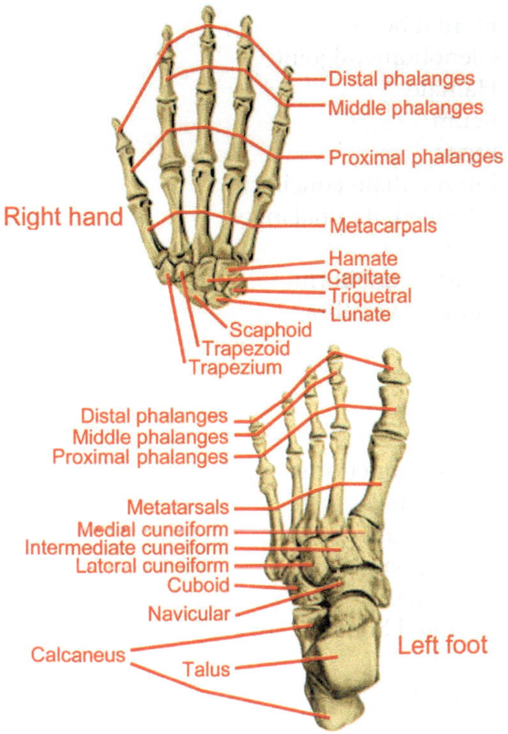

FIGURE 2.1.2 Hands and feet.

flat, C-shaped bone. Five sets of Ribs are known as "false ribs," three of these sharing a common cartilaginous connection to the sternum, while the last two (11th and 12th ribs) are termed floating ribs. They are attached to the vertebrae only, and not to the sternum.

The pelvis is that wide left and right bone that forms our hips. It is that bony structure that connects the base of the spine or sacrum and coccyx to the upper end of the rear legs. Anatomically, the pelvis can be split into three regions: the Ilium, the wide or hip regions, the Pubis or groin regions, and the Ischium including the femur joint. The Pelvis stretches from the wide left and right hip wings, the Iliac Crests, to the lower front where the two sides of the pelvis meet at the Pubic Synthesis, behind the vagina or penis.

It is important to know and understand the anatomy of the body's major bone joints. The largest joint is where the thighs or femurs meet the pelvis or the hip joint. This joint allows humans to walk, run, and jump. This joint bears our body's weight and the force of some of our strongest muscles. It is a round, cup-shaped ball and socket synovial joint and is called the acetabulum.

The most flexible joint in body is the shoulder joint, formed by the union of the humerus, scapula, and clavicle. Commonly thought of as a single joint, it is actually made up of two separate joints, the glenohumeral and acromioclavicular joints. These joints work together allowing the arm both to move in a large circle and to rotate around its axis at the shoulder. The glenohumeral joint is a ball-and-socket joint formed between the rounded head of the humerus and the cup-like depression of the scapula. The acromioclavicular joint is formed by an articulation between the ends of the clavicle and the scapula. It is a flat gliding joint that gives the shoulder additional flexibility.

The knee is one of the largest, strongest, and most complex joints in the body. It is a synovial hinge type joint. The knee joins the femur to the shin bone or tibia. The smaller bone that runs alongside the tibia or fibula and the kneecap or patella are the other bones that comprise the knee joint. The knee joint is called the tibiofemoral joint.

Careful examination of the skull shows that the cranium is in fact multiple bones crudely joined together to make a brain cavity (Figure 2.1.3). There is in fact the frontal bone, parietal bone, occipital bone, temporal bone, nasal bone, and zygomatic bone aside the maxilla and mandible (Figure 2.1.3).

It is also important to remember that the skull includes three tiny ear bones in the middle ear together called the ossicles that are part of the hearing system in human (see Figure 3.5.3). These are the malleus of hammer bone, the incus, or block that is shaped by the hammer, and the stapes or strapping bone (Figure 3.5.3). Basically, in the hearing process the tiny bone Malleus vibrates with sound against the tiny

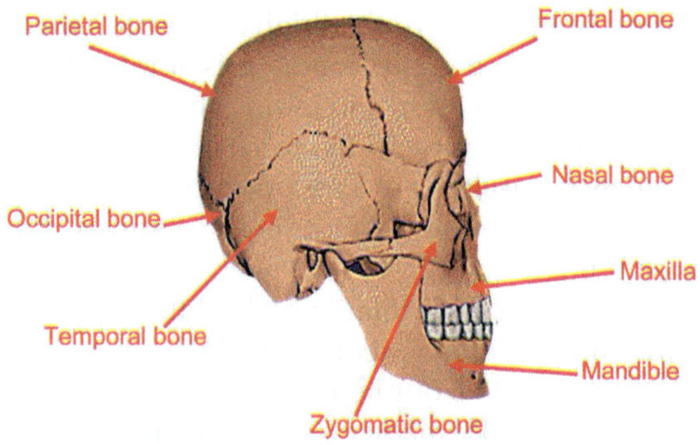

FIGURE 2.1.3 The skull.

bone Incus, and the tiny bone Stapes wedges everything in place. The ossicles mechanically convert the vibrations of the eardrum into amplified pressure waves in the fluid of the inner ear. The sound pressure is oscillated and concentrated by the bones.

CHAPTER KEYWORDS

Know these keywords and what each word refers to as well as its function. Read over the list. Are all these keywords familiar?

Acetabulum
Acromioclavicular joint
Atlas
Axis
Calcaneus
Capitate
Carpals
Cervical vertebra
Clavicle
Cranium
Cuboid
Distal phalanges
False rib
Femur
Fibula
Frontal bone
Glenohumeral joint
Hamate
Ileum
Incus
Intermediate cuneiform
Intermediate phalanges
Ischium
Lateral cuneiform
Lumbar vertebra
Lunate
Malleus
Mandible
Maxilla
Medial cuneiform
Metacarpals
Metatarsals
Nasal bone
Navicular
Occipital bone
Ossicles
Parietal bone
Patella
Proximal phalanges
Pubis
Radius
Ribs
Sacral vertebra
Scaphoid
Scapula
Stapes
Sternum
Talus
Tarsals
Temporal bone
Thoracic vertebra
Tibia
Tibiofemoral joint
Trapezium
Trapezoid
Triquetral
Ulna
Zygomatic bone

INTERNET REFERENCES

Use these references to find out more.
http://en.wikipedia.org/wiki/Skeleton
http://en.wikipedia.org/wiki/Skull
http://en.wikipedia.org/wiki/Human_skeleton
http://www.google.com/search?q=human+bones&client=firefoxa&hs=Own&rls=org.mozilla:en
http://en.wikipedia.org/wiki/List_of_bones_of_the_human_skeleton
http://en.wikipedia.org/wiki/Acetabulum
http://www.innerbody.com/image/skel17.html
http://en.wikipedia.org/wiki/Shoulder
http://www.innerbody.com/image/skel15.html
http://www.webmd.com/pain-management/knee-pain/picture-of-the-knee
http://www.cliffsnotes.com/sciences/anatomy-and-physiology/the-skeletal-system/skull-cranium-and-facial-bones
http://www.ivy-rose.co.uk/HumanBody/Skeletal/Bones_CranialandFacial.php

CHAPTER QUIZ

Answer these questions. Do you know the answers, holy cow! Check yourself out in this quiz, and prove that you are a humanology wiz.

1. Phalanges are found in which two places?
   - **a.** On the skull and on the ribs
   - **b.** Part of the pelvic bone and part of the hand finger bones
   - **c.** The finger bones and the toe bones
   - **d.** Part of the skull and part of the spine
   - **e.** On the big toe and in other toes
2. The collar bone is the
   - **a.** Clavicle
   - **b.** Scapula
   - **c.** Humerus
   - **d.** Cervical disk
   - **e.** Mandible
3. The tibia is found in the
   - **a.** Thigh
   - **b.** Leg
   - **c.** Forearm
   - **d.** Shoulder
   - **e.** Upper arm
4. The acetabulum is the
   - **a.** Shoulder joint
   - **b.** Hip joint
   - **c.** Knee joint
   - **d.** Neck
   - **e.** Cranium
5. The incus bone is part of the
   - **a.** Knee
   - **b.** Hip joint
   - **c.** Knee joint
   - **d.** Neck
   - **e.** Ear

CHAPTER

2.2

Skin, Ligaments, and Muscles

SKIN

The skin is the biggest organ in the body. It offers the organs and muscles of the body containment and protection. The skin is part of the integumentary system made up of multiple layers of ectodermal tissue that protects the underlying muscles, bones, and internal organs in the body.

Humans have hair on their skin. The skin interfaces with the environment and is the first line of defense from external factors protecting the body against pathogens, excessive water loss, temperature regulation, and sensation. The thickness of skin very much varies from bodily location to location. The skin located under the eyes is the thinnest skin in the body at 0.5 mm thick. The skin on the palms and the soles of the feet by comparison is 4 mm thick and is the thickest skin in the body.

Skin comprises an epidermis or protective outer layer, a dermis or soft under layer, and a hypodermis, which comprises fat and connective tissue (Figure 2.2.1). The epidermis is the body's protective barrier responsible for keeping water in the body and preventing pathogens from entering. The epidermis is a sheet-like epithelium. Among other functions, the epidermis helps the skin regulate body temperature.

The dermis is the layer of skin beneath the epidermis and cushions the body from stress and strain. The dermis provides elasticity to the skin through an extracellular matrix composed of collagen fibrils, microfibrils, and elastic fibers. The dermis has many nerve endings that provide the sense of touch and heat in the skin. It also contains the hair follicles, sweat glands, sebaceous glands, and the blood vessels, which provide nourishment for the epidermis.

The hypodermis lies below the dermis (Figure 2.2.1). Its function is to attach the skin to bone and muscle as well as supply it with blood vessels and nerves. Fat in the hypodermis serves as further padding and insulation for the body (Figure 2.2.1).

LIGAMENTS AND CONNECTIVE TISSUE

A ligament is the fibrous tissue or connective tissue that connects bones to other bones. It is also the connective tissue of the body. Ligaments are similar to tendons as they are all made of connective tissue.

Human Physiology, Biochemistry and Basic Medicine
http://dx.doi.org/10.1016/B978-0-12-803699-0.00022-0

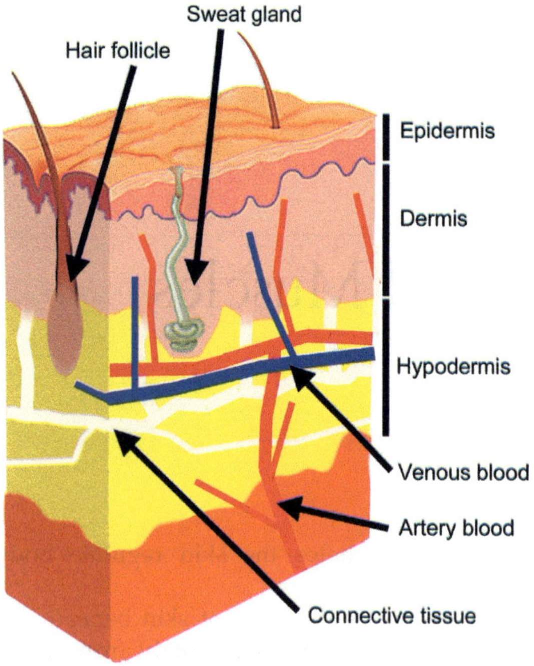

FIGURE 2.2.1 Structure of skin.

Connective tissue is a group of tissues that maintain the form of the body and provide cohesion and internal support. They include several types of fibrous tissues that vary in their density and cellularity, as well as more specialized variants. Some organs are suspended from the wall of a cavity by thin sheets of connective tissue; others are embedded in fat tissue. The entire body is supported by a skeleton composed of bone and a type of connective tissue. The individual bones of the skeleton are held together by ligaments, and muscles are attached to bone by tendons, both of which are forms of connective tissue in which many fiber bundles provide great tensile strength. At joints, the bones are covered with cartilage, a connective tissue with an abundant intercellular substance that gives it a firm consistency. The synovial membrane, which lines joint cavities and lubricates and nourishes joint surfaces, is also a form of connective tissue.

Blood vessels, both large and small, pass through connective tissue, which is closely associated with the nourishment of tissues and organs throughout the body. All nutrient materials and waste products traveling between the organs and the blood must traverse perivascular spaces occupied by connective tissue. One of the important functions of the connective-tissue cells is to maintain conditions in these spaces that favor this exchange.

All forms of connective tissue are composed fibers, an amorphous matrix called ground substance, and stationary and migrating cells. The proportions of these components vary from one part of the body to another depending on the local structural requirements. In some areas, the connective tissue is loosely organized and highly cellular; in others, its fibrous components predominate; and in still others, the ground substance may be its most conspicuous feature. The anatomical classification of the various types of connective tissue is based largely upon the abundance and arrangement of these components.

The fibrous components of connective tissue are of three kinds: collagenous, elastic, and reticular. Most common are the fibers composed of collagen. These fibrous components can be shown under the microscope to be made up of bundles of smaller fibers. The smallest fibers visible with the light microscope can be shown with the electron microscope to be composed of multiple fibrils up to 1000 angstroms (1 angstrom $= 1 \times 10^{-7}$ mm) in diameter. Collagenous fibers are found in tendons, ligaments, skin, and bone.

Elastic fibers allow arteries and organs like the lung to fold and shrink. These are commonly part of an extracellular matrix. An extracellular matrix is a collection of extracellular molecules secreted by cells that provides structural and chemical support to surrounding cells. Reticular fibers form a scaffolding to support other cells. These are also made of the structural protein collagen. Reticular fibers are found in liver, bone marrow, and lymph nodes.

MUSCLES

Muscles are the structural elements that drive human bodily movement, whether the beating of the heart, the raising of the hand, or the exertion of the legs or feet. Muscle tissue varies with function and location through the body. There are three classes of muscle tissue in the body: skeletal or striated muscle, smooth or nonstriated muscle, and cardiac muscle.

Smooth or nonstriated muscle and cardiac muscle move involuntarily, with no conscious interference. These muscle types may be directly controlled by the central nervous system or by receiving hormonal signals. Skeletal muscle or striated muscle only moves consciously or voluntarily upon nervous signal through the central nervous system. Reflexes are a form of nonconscious activation of skeletal muscles.

Muscle cells (myocytes) are elongated cells ranging from several millimeters to about 10 cm in length and are very narrow ranging from 0.01 to 0.1 mm in width. Muscle cells are joined together in tissues that may be either striated or grooved, or smooth, depending on the presence of myofibrillar contractile proteins called myofilaments.

Skeletal muscle is always striated or grooved in structure with myofibrillar contractile proteins. It is anchored by tendons linked to bone and is used to effect skeletal movement such as motion. The average adult male has 42% skeletal muscle by body mass, while the average adult female has 36% skeletal muscle by body mass.

Smooth muscle is found within the walls of organs and structures such as the esophagus, stomach, intestines, bronchi, uterus, urethra, bladder, and blood vessels. Cardiac muscle is found only in the heart. It is a striated or grooved muscle very similar to skeletal muscle. The biggest difference is that it is not subject to voluntary control.

Cardiac and skeletal muscles are packed together into bundles. While skeletal muscles are arranged in parallel bundles, cardiac muscle connects at irregular angles. Striated muscle contracts and relaxes in short, intense bursts, whereas smooth muscle sustains longer or even near-permanent contractions.

Skeletal muscle contraction may be separated into concentric, eccentric, isomeric, and isotonic contractions. Concentric contraction occurs in muscles that literally shorten with contraction. Eccentric contraction occurs in muscles in which the load is lowered gently with contraction, rather than letting it drop. Isomeric contraction is more like a twitch in that the muscle length does not change with contraction. Isotonic contraction is that in which the tension of the muscle remains constant.

The sliding filament theory describes the process whereby all muscles contract (Figure 2.2.2). Muscles contain thin filaments made from the protein Actin and thick filaments made from the protein Myosin. These filaments slide together as illustrated in Figure 2.2.2.

The muscle action starts in the human central nervous system. An electrical signal is sent down the nerve or axon to the neuromuscular junction. The signal causes a calcium ion influx through voltage-gated calcium channels. This leads to acetylcholine release. The acetylcholine release causes a sodium influx. This action spreads through the muscle fibers depolarizing them. This activates the actin containing thin filaments. ATP binds to myosin on the thick filaments providing the energy needed for muscle movement.

A sarcomere (Figure 2.2.2) is the basic unit of a muscle. The Z Disc (Figure 2.2.2) defines the boundaries of a sarcomere or the basic muscle unit. A muscle such as the biceps may contain 100,000 sarcomeres or basic muscle units.

KEY MUSCLES

Several muscles in the body can be defined as key muscles, or the muscles that let us lift our hands in the air, walk, and shake our head. You should know these muscles.

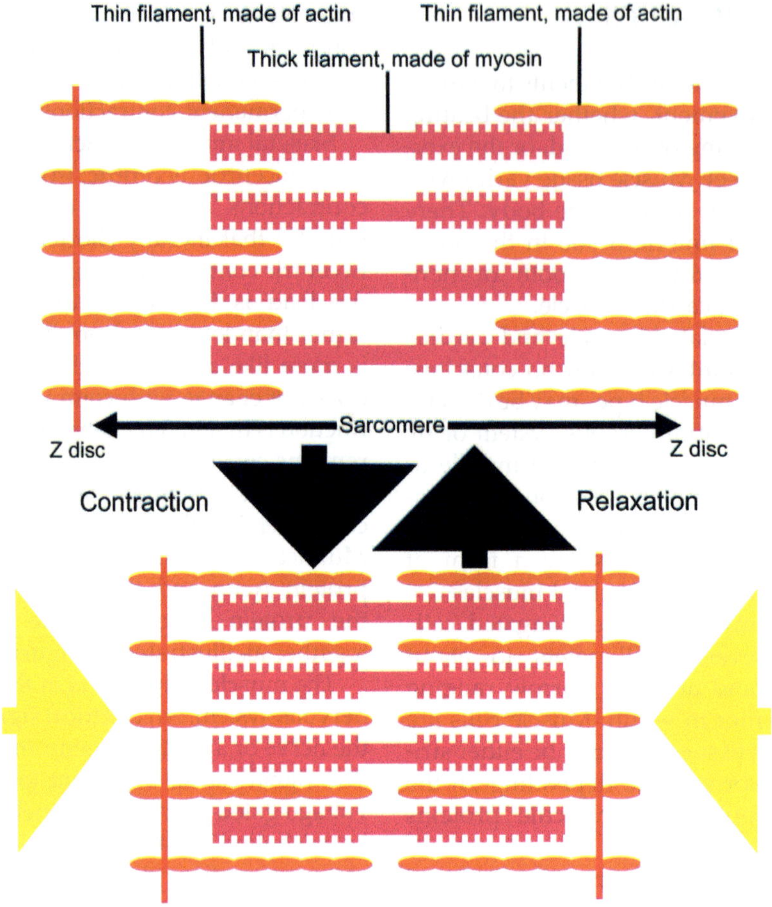

FIGURE 2.2.2 Muscle contraction and relaxation.

There is the Brachialis, which governs flexing the elbow. This is a muscle joined between the humerus bone of the upper arm and the ulna bone on the arm. Above the Brachialis is the giant Biceps Brachii muscle that aids in flexing the arm and principally acts to permit the arm to twist or a screwdriver like action. This muscle is attached at top with its two heads to the scapula bone and attached at the bottom with its two heads to both ulna and radius arm bones. In arm wrestling or Bras de fer, both the Biceps Brochii and Brachialis muscles may be challenged.

The biggest muscle in the leg is Quadriceps Fermoris. This muscle is at the center of kicking actions, walking, and running. This muscle has four heads. Three heads are attached to the femur bone and one is attached to the hip. All four heads lead to the patella or knee bone with attachments extending to the tibia or leg bone.

In the neck, a left side and right side Sternocleidomastoid muscle permits head shaking from left to right. These are attached to the mastoid part of the temporal bone of the cranium and extend down to the clavicle or collarbone.

Gluteus Maximus is a massive muscle, the strongest muscle in the body that forms on the bottom or buttocks. It functions to extend the acetabular joint and brings the bent thigh into a line with the body. Its most powerful action is to cause the body to regain the erect position after stooping, by drawing the pelvis backward.

CHAPTER KEYWORDS

Know these keywords and what each word refers to as well as its function. Read over the list. Are all these keywords familiar?

Acetylcholine
Actin
Biceps Brachii
Brachialis
Cardiac muscle
Cartilage
Collagen
Collagenous fibers
Concentric contraction
Connective tissue
Dermis
Eccentric contraction
Ectodermal
Elastic fibers
Epidermis
Extracellular matrix
Extracellular fiber
Gluteus Maximus
Ground substance
Hypodermis
Isomeric contraction
Isotonic contraction
Ligament
Myocytes
Myofibrillar
Myofilament
Myosin
Neuromuscular junction
Perivascular space
Quadriceps Femoris
Reticular Fibers
Sarcomere
Skeletal muscle
Smooth muscle
Sternocleidomastoid
Striated
Striated muscle
Synovial membrane
Tendon

INTERNET REFERENCES

Use these references to find out more.
www.webmd.com/skin-problems-and-treatments/picture-of-the-skin
en.wikipedia.org/wiki/Skin
en.wikipedia.org/wiki/Dermis
dermatology.about.com/cs/skinanatomy/a/anatomy.htm
en.wikipedia.org/wiki/Connective_tissue
biology.about.com/od/anatomy/a/aa122807a.htm
www.britannica.com/EBchecked/topic/132995/connective-tissue
en.wikipedia.org/wiki/Muscle
en.wikipedia.org/wiki/Muscle_tissue
en.wikipedia.org/wiki/Muscle_contraction
www.ncbi.nlm.nih.gov/pubmed/21314430
http://en.wikipedia.org/wiki/Myosin

CHAPTER QUIZ

Answer these questions. Do you know the answers, holy cow! Check yourself out in this quiz, and prove that you are a humanology wiz.

1. The three components of the skin are:
   a. Myodermis, epidermis, and hyperdermis
   b. Epidermis, collagenous layer, and hyperdermis
   c. Epidermis, episquamous layer, and hyperdermis

   d. Dermis, hypodermis, and hyperdermis
   e. Epidermis, dermis, and hypodermis
2. What function does fat in the skin serve?
   a. Provides the energy needed for skin movements
   b. Provides the energy needed for skin insulation
   c. Provides energy for adjacent muscles
   d. Provides padding and insulation
   e. Shields blood vessels approaching the surface of the skin
3. Which are the correct functions of connective tissues?
   a. Tendons, cartilage at joints, synovial membranes, ligaments, suspends organs
   b. Tendons, synovial membranes, fibrous joints, ligaments, heads of muscles
   c. Tendons, suspends organs, trachea, urethra, ligaments
   d. Tendons, suspends organs, muscle filaments, ligaments
4. The fibrous components of connective tissue are which three kinds?
   a. Collagenous, elastic, and reticular
   b. Collagenous, flexible, and antiflexable
   c. Collagenous, anticollagenous, and flexible
   d. Collagenous, flexible, and reticular
   e. Collagenous, elastic, and antielastic
5. There are three clear types of muscles in the body:
   a. Functional or striated muscles, smooth or nonstriated muscle, and cardiac muscle
   b. Skeletal or striated muscles, smooth or nonstriated muscle, and cardiac muscle
   c. Essential or striated muscles, smooth or nonstriated muscle, and cardiac muscle
   d. Skeletal or nonstriated muscles, smooth or striated muscle, and vascular muscle
   e. Skeletal or nonstriated muscles, smooth or striated muscle, and cardiac muscle

CHAPTER

2.3

Apoptosis, Growth, and Aging

Apoptosis is the process of programmed cell death that occurs in all multicellular organisms. Organic events, bulges and shrinkage in cells, nuclear fragmentation, chromatin and DNA fragmentation, are all part of apoptosis process and the cell changes and leads to cell death (Figure 2.3.1). Cell death or apoptosis leads to cell replication or cell replacement. Apoptosis confers advantages to cellular life cycle. For example, the separation of fingers and toes in human embryos occurs because these cells apoptose. In humans, 50-70 billion cells die each day due to apoptosis. As shown by today's research, excessive apoptosis causes atrophy, whereas insufficient apoptosis results in uncontrolled cell growth, like cancer.

All the cells in the human body undergo apoptosis or replace themselves at different rates. The mitochondria of cells control apoptosis. Mitochondria produce SMACs (small mitochondria-derived activator of caspases). These are released into the cellular cytoplasm where they bind and inhibit apoptosis proteins (IAPs) and deactivate them, preventing the IAP from blocking the apoptosis and therefore also allowing it to proceed.

After a cell receives the IAP stimulus, it undergoes organized degradation of cellular organelles by activated proteolytic caspases, resulting in cell shrinkage and the organelles becoming tightly packed. Chromatin I in the nucleus becomes condensed and the DNA becomes fragmented. The cell breaks apart into several vesicles called apoptotic bodies (Figure 2.3.1). Apoptosis progresses quickly and its products are quickly lysed and removed (Figure 2.3.1).

The process of apoptosis is critical to cell age. During aging, too little apoptosis in cells stretches the life of cells, over long survival limits cells age or poorly function. The loss of muscles and strength with aging, for instance, is a prevalent among older adults. Evidence suggests that the age-related loss due to decreasing apoptosis is the mechanism driving muscle loss in old people and loss of strength.

Telomeres are a region of repetitive DNA nucleotide sequences at each end of a chromosome. In humans, the DNA sequence of nucleotides in telomeres is TTAGGG. Telomeres have been shown to shorten with each successive cell division or cell replacement or apoptosis (Figure 2.3.2). Shortened telomeres seemingly prevent cell multiplication or apoptosis. This shortening process may cause cellular aging by blocking cell growth and apoptosis (Figure 2.3.2).

Human Physiology, Biochemistry and Basic Medicine
http://dx.doi.org/10.1016/B978-0-12-803699-0.00023-2

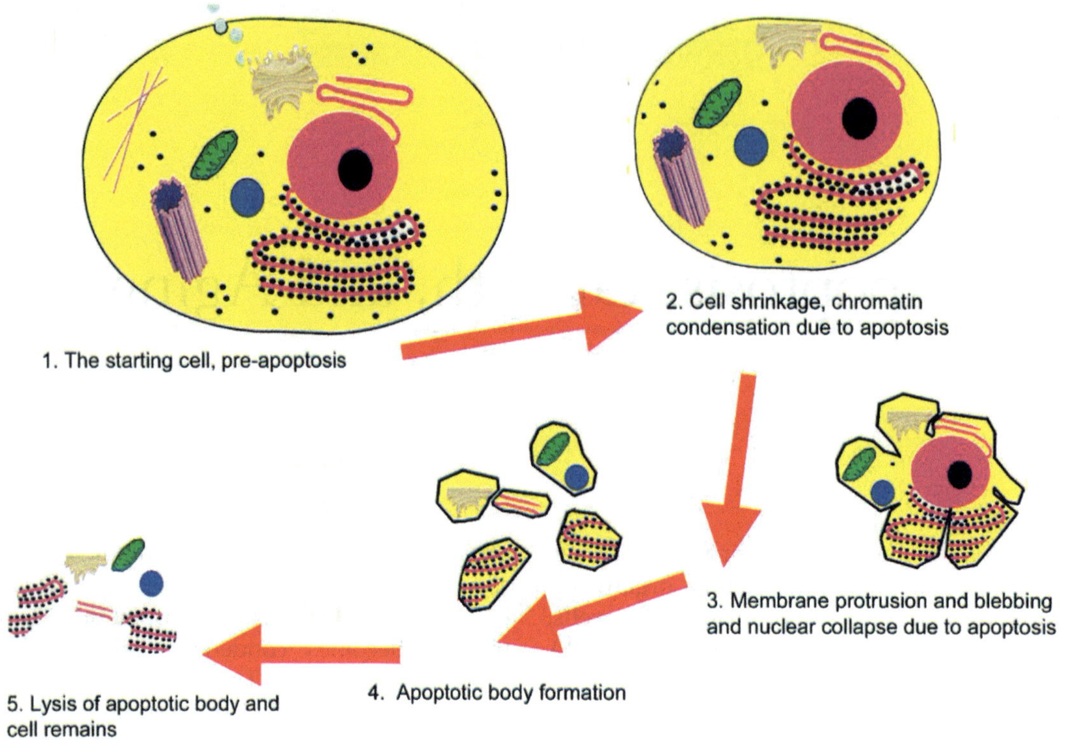

FIGURE 2.3.1 Apoptosis.

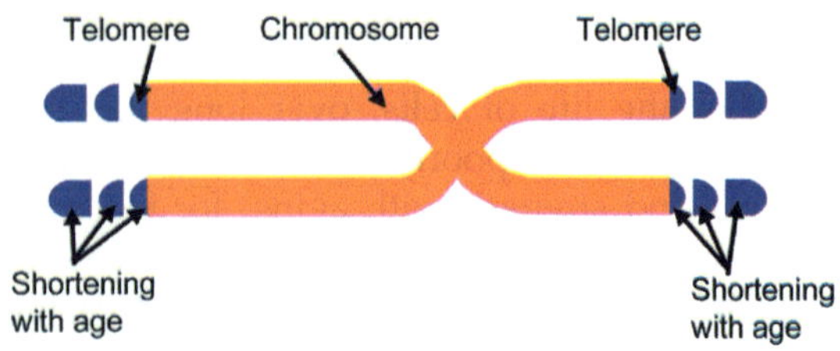

FIGURE 2.3.2 Telomeres and aging.

In laboratory settings, researchers have demonstrated that selected alterations in specific genes can extend life span quite substantially in fruit flies and possibly mice. Life-span extension can occur as the result of genetic alterations that increase DNA repair and reduce oxidative damage in species.

Many other mechanisms of aging have been proposed. According to the reproductive-cell cycle theory, aging may be regulated by reproductive hormones that regulate the cell cycle. According to the DNA damage theory, DNA damage causes a decrease in cell activity, or induced apoptosis. The gene theory is based on loss of DNA by cells each year. The autoimmune theory shows an increase in autoantibodies that attack the body's tissues as the body ages. According to the misrepair theory, aging is the result of the accumulation of "misrepair," or an inability to repair cells with aging. Much more research is needed to confirm the telomere and apoptosis theories of aging.

When cells begin to become cancerous, they divides more often and the telomeres become very short. If the telomeres become too short, the cells may die. Cells escape this fate by becoming a cancer cell and activating an enzyme called telomerase, which prevents the telomeres from getting even shorter. Research has found shortened telomeres in many cancers, including pancreatic, bone, prostate, bladder, lung, kidney, and head and neck cancers.

Cancer cells require a mechanism to maintain their telomeric DNA in order to continue dividing indefinitely. Telomerase is the natural enzyme that promotes telomere repair. It is active in stem cells, germ cells, and in over 90% of cancer cells. Telomerase functions by adding bases to the ends of the telomeres. As a result of telomerase activity, these cells seem to possess a kind of immortality.

Recent discoveries show that apoptosis and the genes that control apoptosis have major roles in human cancers. Carcinogenesis may override apoptosis permitting cancer cell growth rather than failure. Finally, it is now documented that most anticancer agents induce apoptosis. An intense research effort is uncovering the underlying mechanisms of apoptosis, including new cancer treatment strategies to exploit apoptosis for therapeutic benefit. Both telomeres and apoptosis may be critical cancer. Their aging action may induce cancer in inefficient cells in older people. Telomerase and telomere extension and blockage of apoptosis may promote cancers and advanced disease.

CHAPTER KEYWORDS

Know these keywords and what each word refers to as well as its function. Read over the list. Are all these keywords familiar?

Apoptosis
Autoimmune
Blebbing
Caspase
Chromatin
DNA damage
IAP
Inhibit apoptosis protein
Misrepair
Nuclear collapse
Nuclear fragmentation
SMAC
Small mitochondria-derived activator of caspases
Telomerase
Telomere
Carcinogenesis

INTERNET REFERENCES

Use these references to find out more.
en.wikipedia.org/wiki/Apoptosis
en.wikipedia.org/wiki/Telomere
www.ncbi.nlm.nih.gov/books/NBK26873/
users.rcn.com/jkimball.ma.ultranet/BiologyPages/A/Apoptosis.html
learn.genetics.utah.edu/content/chromosomes/telomeres/www4.utsouthwestern.edu/cellbio/shay-wright/intro/facts/sw_facts
www.news-medical.net/health/Telomeres-and-Cancer.aspx
hmg.oxfordjournals.org/content/10/7/677.full
carcin.oxfordjournals.org/content/21/3/485.full
www.jeccr.com/content/30/1/87

CHAPTER QUIZ

Answer these questions. Do you know the answers, holy cow! Check yourself out in this quiz, and prove that you are a humanology wiz.

1. Apoptosis is
- **a.** Programmed cell death
- **b.** Programmed cell replication
- **c.** Programmed destruction of cell mitochondria
- **d.** Programmed destruction of cell nucleus
- **e.** A cell destroying an adjacent cell

2. Apoptosis is controlled by
   a. A cell nuclear protein, IAP
   b. A cell mitochondria protein, IAP
   c. A cell lysis protein IAP, made by lysosomes
   d. A circulating hormone called IAP
   e. A protein made by endoplasmic reticulum called IAP
3. Telomeres are
   a. Repeating DNA sequences located at the centromeres of chromosomes
   b. Repeating DNA sequences located at the 4 ends of chromosomes
   c. Polymers of A located at the centromeres of chromosome
   d. Polymers of A located at the 4 ends of chromosomes
   e. CCG sequences located at the centromeres of chromosomes
4. As cells age and go through numerous divisions telomeres become
   a. Longer
   b. Wider
   c. More repetitive
   d. More active
   e. Shorter

Section 3
Human Organs

CHAPTER

3.1

The Endocrine System

The endocrine system or hormone system partly ties the body and all its organs together. Figure 3.1.1 illustrates the major endocrine pathways controlling the human body. An endocrine or hormone is a molecule produced by an endocrine gland that controls function of distant organs by binding a specific receptor protein on the distant organ. This chapter deals with the major human endocrine systems. The reproductive endocrines are discussed in Chapters 3.8–3.10. Multiple other minor or support endocrine systems exist. There are also multiple other molecules that bind receptors that are organ control signals; these include the large folded lipids or fatty acids called prostaglandins, and systems like the renin-angiotensin system, which controls blood vessel constriction.

HYPOTHALAMUS-ANTERIOR PITUITARY AXIS

The hypothalamus is a small part of the brain and the pituitary gland is an endocrine organ directly below the brain (see Chapter 3.4, Figure 3.4.5). If you place your finger at the back of the roof of your mouth, the pituitary can be considered as one inch above the finger, and the hypothalamus of the brain one inch above the pituitary. While the hypothalamus is brain tissue and not endocrine tissue, the hypothalamus does deposit four neurohormones into the portal or local circulation that links the hypothalamus and anterior pituitary gland, the hypothalamic hypophyseal portal circulation. The pituitary is divided into the anterior pituitary, which secretes hormones in response to hypothalamic endocrine signals and the posterior pituitary, which is mostly composed of hypothalamic nervous tissue extensions.

The hypothalamus of the brain produces gonadotropin-releasing hormone (GnRH), thyrotropin-releasing hormone (TRH), corticotropin-releasing hormone (CRH), growth hormone-releasing hormone (GHRH), and prolactin-releasing hormone (PRLH). These hormones pass down the hypothalamic hypophyseal portal circulation to thyrotrope cells, corticotropic cells, somatotropic cells, and lactotropic cells of the anterior pituitary gland. These five releasing factors are small peptide derivatives. For instance, GnRH is a decapeptide or 10 amino acid molecule modified into a peptide amide (has a covalent NH_2 group covalently added to C-terminus). GnRH acting on anterior pituitary gonadotrope cells promotes the

Human Physiology, Biochemistry and Basic Medicine
http://dx.doi.org/10.1016/B978-0-12-803699-0.00003-7

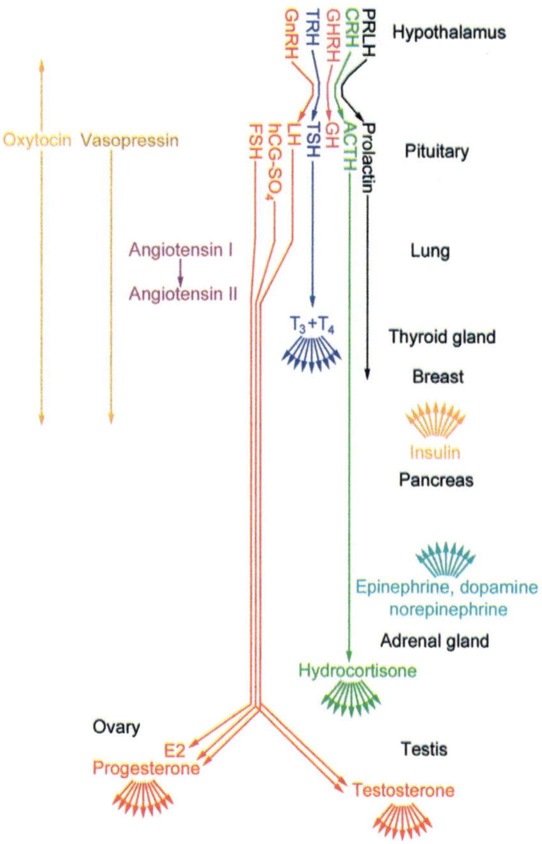

FIGURE 3.1.1 The endocrine system marking the main human hormones.

production of luteinizing hormone (LH), sulfated human chorionic gonadotropin (hCG-SO_4), and follicle stimulating hormone (FSH). Three large dimeric glycoprotein hormones or glycopeptins that share a common 92 amino acid α-subunit have an independent ß-subunit of 121 to 145 amino acids. TRH acts on anterior pituitary thyrotrope cells releasing thyroid stimulating hormone (TSH), as well as a glycoprotein hormone sharing the common 92 amino acid α-subunit. CRH acts on anterior pituitary corticotropic cells producing the protein adrenocorticotropic hormone (ACTH). ACTH is a 39 amino acid single peptide protein. GHRH acts on anterior pituitary somatotropic cells produced by the protein growth hormone (GH) and PRLH actin on lactotropic cells causes the release of prolactin.

All these hypothalamic releasing hormones and anterior pituitary hormones generate smooth operating feedback cycles that balance hormone production. GnRH promotes LH, hCG-SO_4, and FSH pituitary production. These hormones in turn circulate through the female blood and promote ovarian estradiol (E2) production. Estradiol circulates through the blood and modulates hypothalamic GnRH pulses. In men, LH and hCG-SO_4 control testosterone production. Testosterone feeds back and modulates GnRH pulses. Similarly, TRH promotes TSH production. TSH circulates and acts on thyroid gland thyroxine (T_4) and triiodothyronine (T_3) production. T_4 in turn feeds back and modulates TRH pulses. CRH promotes ACTH production. ACTH circulates in the body and acts on the adrenal gland to promote hydrocortisone production. Hydrocortisone feeds back and modulates CRH pulses. The pathway is simpler for GH and prolactin. GHRH promotes GH production. GH directly feeds back and modulates GHRH pulses. Similarly PTLH promotes prolactin production, which feeds back and modulates PRLH pulses.

GH is produced by somatotropic cells and is the only pituitary final hormone product. It is a large protein of 191 amino acids. GH is anabolic endocrine promoting all around growth in children and adolescents. It increases bone growth and Ca retention. It increases bodily muscle mass, bodily protein synthesis, and uptake of glucose.

Laurence Cole, PhD had his pituitary removed 25 years ago because of a large tumor extending into the brain. This led to him having to take multiple replacement hormones. For 8 years from age 45 to 53 he gave himself daily injections of GH. He found that these endocrine injections were seemingly doing nothing to his adult body. He stopped taking the shots and has noticed no illicit effects. In this respect, GH can be called strictly an essential hormone produced in children and adolescents.

Prolactin is a protein that in humans is most known for its function in females promoting milk production. Its influence, however, over a much larger number of functions has been demonstrated. Prolactin stimulates the mammary glands to produce milk after birth. Serum concentrations of prolactin increase during pregnancy causing the enlargement of the mammary glands of the breasts in preparation for the production of milk.

Prolactin also provides the body with sexual gratification after intercourse acts. The hormone counteracts the effect of dopamine, responsible for sexual arousal. This is thought to cause the sexual obstinate period. The quantity of prolactin produced can be an indicator of the amount of sexual satisfaction and relaxation. Highly elevated levels of prolactin decrease the circulating concentration of estrogen in women and testosterone in men.

Prolactin also has other functions. It stimulates the growth of oligodendrocyte precursor cells. These cells are responsible for the formation of myelin coatings on axons of neurons in the central nervous system. Prolactin also promotes pulmonary surfactant synthesis during pregnancy, causing the accumulation of essential surfactant in fetal lungs at the end of the pregnancy.

POSTERIOR PITUITARY

The posterior portion of the pituitary gland is largely a collection of axonal projections from the hypothalamus that terminate behind the anterior pituitary. These cells store and produce the protein hormones vasopressin, also known as anti-diuretic hormone (ADH), and oxytocin. These hormones are both nine amino acid peptide amides (with a covalent NH_2 group covalently added to C-terminus). Both hormones are made in the hypothalamus of the brain and are stored in the posterior pituitary cells and released into the circulation.

Vasopressin functions are to control retention of water by the body. Vasopressin promotes water permeability in the kidneys, collecting ducts and distal tubules by inducing translocation of aquaporin-CD water channels in the kidney collecting duct plasma membranes. It plays a major role in blood homeostasis by modulating blood water and salt balance.

Oxytocin has multiple functions in men and women. It primarily acts in women during labor and delivery and during maternal lactation. It stimulates uterine contractions, which are important for cervical dilation for childbirth. It also promotes lactation or breastfeeding. It causes milk to enter subareolar sinuses for excretion through nipples. Research is now proving other critical functions of oxytocin production. When a person hugs or kisses a loved one oxytocin levels rise. During sexual behavior oxytocin levels rise, and during non-sexual selective social bonding oxytocin levels rise in an individual. Strangely, oxytocin levels may also rise during a time of stress. Oxytocin receptors are found in the male and female brain, and in the female uterus and breast tissue.

THYROID GLAND

TRH promotes thyrotrope cells of the anterior pituitary to produce TSH, which promotes the thyroid gland to make T_3 and T_4. Both hormones are iodinated derivatives of the amino acid tyrosine. The major form of thyroid hormone in the blood is thyroxine (T_4), which has a longer circulating half-life than T_3. The ratio of T_4 to T_3 released into the blood is roughly 20 to 1. T_3 and T_4 are synthesized in the follicular cells of the thyroid gland. As a hormone, T_3 is three- to fivefold more potent than T_4.

As hormones, T_3 and T_4 act on nearly every cell in the body, to increase the basal metabolic

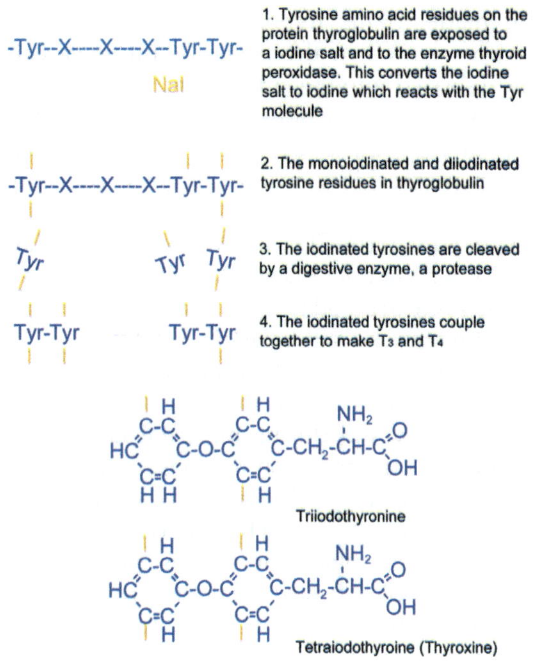

FIGURE 3.1.2 The synthesis of T_3 and T_4.

rate, help control long bone growth, and to increase the body's sensitivity to catecholamines (epinephrine, norepinephrine, and dopamine; see section on Adrenal Medulla). These endocrines also regulate protein, fat, and carbohydrate metabolism, and how human cells use energetic compounds. Thyroid endocrines also lead to heat generation in humans by an unknown mechanism. One effect of administering T_4 to humans is a major drop in body temperature.

Figure 3.1.2 shows how T_3 and T_4 are made in thyroid follicular cells. A protein called thyroglobulin has multiple tyrosine amino acid residues. These are iodinated by the enzyme thyroid peroxidase, an enzymatic reaction that utilizes an iodide salt and hydrogen peroxide. The hydrogen peroxide converts the iodide salt to iodine, which reacts with the tyrosine residues. The iodinated tyrosine in thyroglobulin is cleaved by a protease enzyme releasing the iodo-tyrosine molecules. The monoiodinated and diiodinated iodo-tyrosine molecules are then coupled (monoiodinated + diiodinated = T_3, diiodinated + diiodinated = T_4) by the action of thyroid peroxidase to make T_3 and T_4, the two endocrines (Figure 3.1.2).

ADRENAL CORTEX

The hypothalamic releasing factor CRH promotes anterior pituitary corticotropic cells to produce the protein ACTH. The corticotropic cells secrete a much larger protein called preproopiomelanocortin (241 amino acids long), which is cleaved and broken down to make a 39 amino acid ACTH and several melanocyte management hormones and endorphins, or natural pain and stress management molecules. ACTH acts on the adrenal cortex, which comprises the bulk of the adrenal gland (adrenal medulla is surrounded by the adrenal cortex) to promote production of two steroid endocrines, a mineralocorticoid hormone called aldosterone and a glucocorticoid hormone called hydrocortisone.

Aldosterone circulates throughout the body and acts on an intracellular receptor within the tubules of the kidney. It regulates and activates the Na^+/K^+ balance pumps within the kidney. This creates a concentration gradient, which results in reabsorption of Na^+ ions and water into the blood and the excretion of K^+ ions into the urine.

Hydrocortisone is a human stress hormone. Hydrocortisone elevates the blood pressure and prepares the body for a fight or flight response. When stress is detected, hydrocortisone generates blood glucose to cope with the body's stress issue. Hydrocortisone promotes the human immune system and in that respect has become an administered universal panacea to promote healing.

As shown in Figure 3.1.3, hydrocortisone and aldosterone are steroids produced from the core

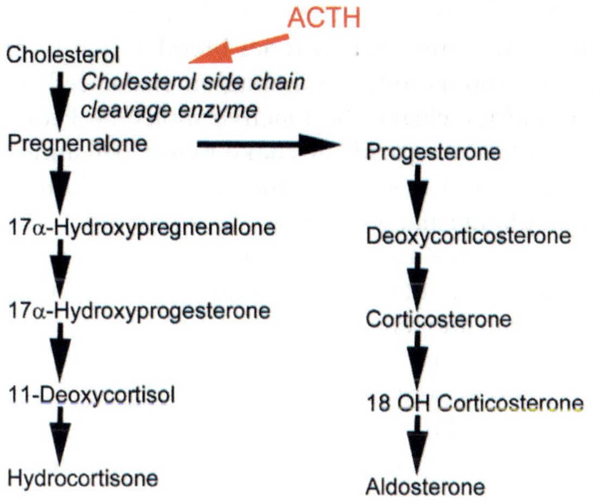

FIGURE 3.1.3 Hydrocortisone and aldosterone synthesis.

steroid cholesterol. Cholesterol is a 27-carbon lipid or fatty acid folded in the steroid style with three fused 6-carbon rings and one 5-carbon ring, and a branched 8-carbon side chain. ACTH promotes cholesterol side chain cleavage enzyme, which removes six of the eight carbons on the side chain. Pregnenalone is converted through four enzymatic steps to the 21-carbon molecules hydrocortisone and aldosterone. These steroid hormones maintain the fused ring architecture of the parent molecule cholesterol.

ADRENAL MEDULLA

Like the hypothalamus of the brain, which secretes hormones, the adrenal medulla comprises mostly nervous tissue. Unlike the adrenal cortex which responds to protein signals to produce hydrocortisone and aldosterone, the adrenal medulla responds to nerve signals to produce catecholamines, three endocrines, epinephrine (also called adrenaline), norepinephrine (noradrenaline), and dopamine. Epinephrine, norepinephrine, and dopamine are molecules that also function as neurotransmitters, molecules that transmit a nerve-to-nerve signal, in the human brain. Like the thyroid hormones T_3 and T_4 epinephrine, norepinephrine and dopamine are synthesized from the amino acid tyrosine.

Epinephrine, norepinephrine, and dopamine raise blood pressure by constricting arteries, promoting breathing, increasing heart beat rate, releasing glucose from energy stores (glycogen), and increasing brain oxygen consumption. In a person with myocardial infarction or heart attack, increased epinephrine is normally seen. Epinephrine, norepinephrine, and dopamine and their inhibitors are common treatments for heart attack and high blood pressure. Epinephrine and norepinephrine are commonly considered as human stress hormones, arousing a person in stress and raising their pulse rate.

The synthesis of these catecholamines (Figure 3.1.4) in the adrenal medulla from tyrosine involves oxygen and the oxygen cofactors biopterin and ascorbate or Vitamin C. In just

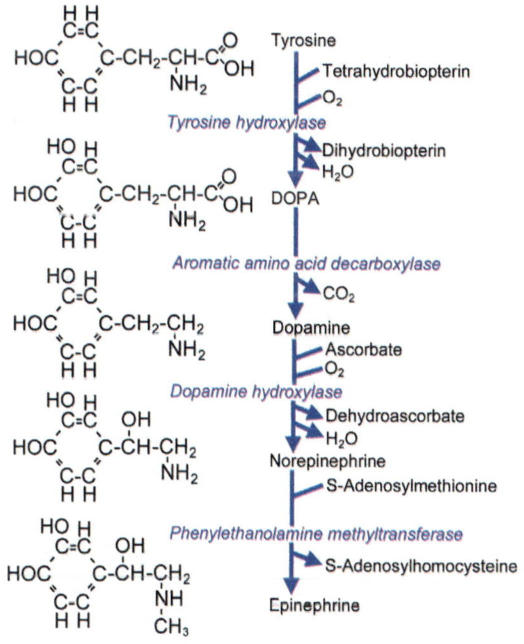

FIGURE 3.1.4 The synthesis of epinephrine, norepinephrine, and dopamine.

two enzymatic steps tyrosine is converted to dopamine and in 3 and 4 enzymatic steps into norepinephrine and epinephrine (Figure 3.1.4).

PANCREATIC ISLETS OF LANGERHANS CELLS

The beta cells of the Islets of Langerhans of the pancreas produce an important endocrine, insulin, which controls human blood glucose levels, glucose, and fat metabolism. Insulin is a protein comprising 51 amino acids, and contains an A and B chain linked together by disulfide bonds. There is no insulin precursor or promoting molecule coming from the pituitary gland; it is strictly produced in response to food or blood glucose. Insulin is produced when any of several stimuli are detected. Stimuli include ingested protein and glucose in the blood produced from digested food. Carbohydrates can be polymers of simple sugars or simple sugars themselves.

Secreted insulin does many things to lower blood glucose content. First, it controls cell uptake of glucose in muscle and adipose or fat tissue. Second, it increases glycogen synthesis in liver and muscle cells (Chapter 1.2). Lower levels of insulin due to low blood glucose, in contrast, cause liver cells to convert glycogen to glucose and excrete it into the blood. Third, it increases lipid synthesis forcing adipose tissue or fat cells to take in blood lipids, which are converted to fatty acids and triglycerides. In contrast, lack of insulin causes the reverse. Fourth, it decreases gluconeogenesis or the production of glucose from non-sugar substrates in the liver. Fifth, it forces cells to absorb circulating amino acids.

RENIN-ANGIOTENSIN SYSTEM

The renin-angiotensin system is a little known hormonal cascade. However, it is probably one of the most important endocrine systems in that it alone controls and balances human blood pressure. When renal blood pressure or blood flow is reduced, juxtaglomerular cells in the kidney cleave the inactive molecule Prorenin, 406 amino acids, to make Renin, 340 amino acids. Renin is secreted directly by the kidney juxtaglomerular cells into the circulation. The liver secretes a large 453 amino acid precursor protein Angiotensinogen into the circulation. Angiotensinogen is converted by the renal enzyme renin into angiotensin I, a small 12 amino acid peptide derived from Angiotensinogen. In the lungs, Angiotensin-converting enzyme (ACE) cleaves Angiotensin I further to make the 10 amino acid hormone Angiotensin II (Figure 3.1.5). Angiotensin II is a potent vasoactive peptide hormone that constricts all blood vessels, increasing blood pressure as needed.

Plasma Renin carries out the conversion of angiotensinogen released by the liver to Angiotensin I. Angiotensin I is subsequently converted to Angiotensin II by the enzyme angiotensin-converting enzyme made in lung pulmonary endothelium cells (Figure 3.1.5). Angiotensin II is a potent vasoactive peptide that causes blood vessels to constrict.

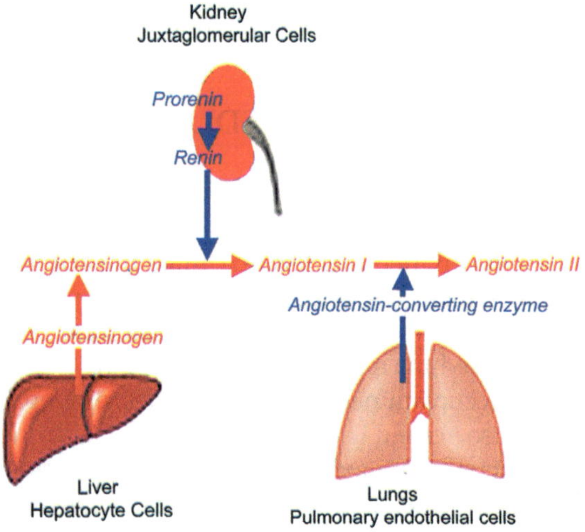

FIGURE 3.1.5 The renin-angiotensin hormone cascade.

Angiotensin II increases blood pressure by promoting Gq protein in the vascular smooth muscle cells. This in turn activates an inositol triphosphate signaling mechanism leading to a rise in intracellular calcium levels and causing contraction.

The hormone angiotensin II also stimulates the secretion of the hormone aldosterone from the adrenal cortex (see Adrenal Cortex). Aldosterone promotes the tubules of the kidneys to increase the reabsorption of sodium and water into blood. This effectively increases the volume of fluid in the body, which also increases blood pressure. Angiotensin II also promotes production of vasopressin by the posterior pituitary and production of ACTH by the anterior pituitary. The renin-angiotensin system balances the body's blood pressure. Angiotensin II receptor inhibitors and angiotensin-converting enzyme inhibitors are the basis for multiple blood pressure drugs.

GONADS

As described in Chapters 3.8 and 3.9, GnRH promotes pituitary gonadotrope cells to make FSH, LH, and hCG-SO_4. These act on gonads, the ovaries in women and testis in men to promote estradiol, an estrogen and progesterone synthesis in women, and testosterone and estradiol synthesis in men. Estradiol and testosterone both feed back to the hypothalamus and balance GnRH production.

CHAPTER KEYWORDS

Know these keywords and what each word refers to as well as its function. Read over the list. Are all these keywords familiar?

11-Deoxycortisol
17α-hydroxypregnenalone
17α-hydroxyprogesterone
ACE
ACE inhibitor
ACTH
Adrenal cortex
Adrenal medulla
Adrenocorticotropic hormone
Aldosterone
Angiotensin I
Angiotensin II
Angiotensin-converting enzyme
Angiotensinogen
Anterior pituitary
Aromatic amino acid decarboxylase
Ascorbate
Catecholamine
Cholesterol
Cholesterol side chain cleavage enzyme
Corticosterone
Corticotrope cell
Corticotropin-releasing hormone
CRH
Deoxycorticosterone
DOPA
Dopamine
Dopamine hydroxylase
E2
Endocrine
Epinephrine
Estradiol
Follicle-stimulating hormone
Follicular cell
FSH
GH
GHRH
GnRH
Gonadotrope cell
Gonadotropin-releasing hormone
Growth hormone
Growth hormone-releasing hormone
hCG-SO_4
Hydrocortisone
Hypothalamic hypophyseal portal circulation
Hypothalamus
Insulin

Iodinated
Islet of Langerhans
Lactotropic cells
LH
Luteinizing hormone
Oxytocin
Norepinephrine
Phenylethanolamine methyltransferase
Pregnenalone
Pre-proopiomelanocortin
Progesterone
Prolactin
Prolactin-releasing hormone
PRLH
Receptor
Renin
Somatotropic cell
Sulfated human chorionic gonadotropin
T_3
T_4
Testosterone
Tetraiodothyronine
Thyroglobulin
Thyroid gland
Thyroid peroxidase
Thyroid stimulating hormone
Thyrotrope cell
Thyrotropin releasing hormone
Thyroxine
TRH
Triiodothyronine
TSH
Tyrosine hydroxylase
Vasopressin

INTERNET REFERENCES

Use these references to find out more.
en.wikipedia.org/wiki/Hypothalamus
www.vivo.colostate.edu/hbooks/pathphys/endocrine/hypopit/anatomy.html
www.inkling.com/read/clinical-gynecologic-endocrinology-infertility-8th/chapter-5/hypothalamic-hypophyseal-portal
en.wikipedia.org/wiki/Pituitary
en.wikipedia.org/wiki/Thyroid
en.wikipedia.org/wiki/Hypophyseal_portal_system
en.wikipedia.org/wiki/Insulin
en.wikipedia.org/wiki/Adrenal_medulla
en.wikipedia.org/wiki/Adrenal_cortex
en.wikipedia.org/wiki/Growth_hormone
en.wikipedia.org/wiki/Renin%E2%80%93angiotensin_system

CHAPTER QUIZ

Answer these questions. Do you know the answers, holy cow! Check yourself out in this quiz, and prove that you are a humanology wiz.

1. Which statement is false regarding the hypothalamic-pituitary axis?
   - **a.** The hypothalamus and anterior pituitary are linked by the hypothalamic-hypophyseal portal circulation.
   - **b.** The hypothalamus as brain tissue secretes at least four pituitary-releasing hormones.
   - **c.** The hypothalamus is linked to the anterior pituitary by nervous tissue.
   - **d.** The hypothalamus releases GnRH, TRH, CRH, and GHRH.
   - **e.** The pituitary is approximately one inch below the brain hypothalamus.
2. In the production of thyroid hormones which statement is true?
   - **a.** The iodination of thyroid hormones involves enzymatic action on the protein thyroglobulin.
   - **b.** The iodination of thyroid hormones involves enzymatic action on the free amino acid tyrosine.
   - **c.** The iodination of thyroid hormones involves the enzyme Tyrosine Oxidase.
   - **d.** The iodination of thyroid hormones involves Potassium Metaperiodate.
   - **e.** The iodination of thyroid hormones involves enzymatic action on tyrosine residues on thyroid stimulating hormone

3. Which statement is false about the endocrine hydrocortisone?
   a. Hydrocortisone is a steroid.
   b. Hydrocortisone is produced by adrenal cortex tissue.
   c. Hydrocortisone stimulates the immune system.
   d. Hydrocortisone contains 21 carbons.
   e. Hydrocortisone is a catecholamine.
4. Which statement is false regarding epinephrine?
   a. Epinephrine is made by the enzymatic oxidation of tyrosine.
   b. Epinephrine is a catecholamine.
   c. Epinephrine is also a brain neurotransmitter.
   d. Epinephrine is sometimes called adrenaline.
   e. Epinephrine is a steroid hormone.
5. Which statement is false regarding insulin?
   a. Insulin is produced by Islets of Langerhans.
   b. Insulin reduces blood glucose.
   c. Insulin is a glycoprotein hormone with a common α-subunit.
   d. Insulin is a protein of 51 amino acids.
   e. Insulin acts to convert glucose to glycogen.

CHAPTER

3.2

The Heart and Lungs

THE HEART

The heart pumps the blood around the body making it a very important organ. The average heart beats approximately 3 billion times in a lifetime. The heart is not heart-shaped; it is a pine cone-shaped organ that is a solid beating muscle (Figure 3.2.1). It weighs approximately 9-12 ounces and measures 5 x 3.5 x 2.5 inches. The vagus nerve is the tenth cranial nerve and interfaces with parasympathetic control of the heart and the digestive tract.

It is the vagus nerve that electrically signals the heartbeat. The vagus nerve follows the carotid artery from the brain through the neck to the heart. Normal sinus rhythm is established by the sinoatrial node (SAN), which is the heart's pacemaker. The SAN is a specialized grouping of cardiomyocyte cells in the upper and back walls of the right atrium of the heart, very close to the heart's input of the superior vena cava. The heart rate of a resting person averages 72 beat per minute. Exercise increases this heart rate.

As shown in Figure 3.2.1, the left ventricle directly feeds the aorta pumping oxygenated blood to the body. The superior and inferior vena cava return the used or deoxygenated blood to the right atrium and then the right ventricle pumps it to the left and right pulmonary arteries (an unusual set of arteries carrying deoxygenized blood, by definition an artery is a vessel transporting blood from the heart) which channels the blood to the lungs. The lungs oxygenate the blood and send it back through the right and left pulmonary veins (by definition a vein carries blood to the heart) to the left atrium of the heart and left ventricle to complete the bodily cycle.

Figure 3.2.2 illustrates this heart-lung body cycle, which shows the four critical heart valves, tricuspid, pulmonary, mitral, and aortic valves, that control blood movement through the heart. The valves are all cusps or flapping points that are pushed open to allow blood to be ejected and which close preventing back flow. Each valve has three or two cusps or flaps. The heart is enclosed in a protective bag, the pericardium, which contains a small amount of fluids. It is located in a compartment in the chest called the mediastinum.

The heart is an active beating organ that has its own blood supply; the root arteries are the right and left coronary arteries. While most blood vessels in the body anastomose to divert

Human Physiology, Biochemistry and Basic Medicine
http://dx.doi.org/10.1016/B978-0-12-803699-0.00025-6

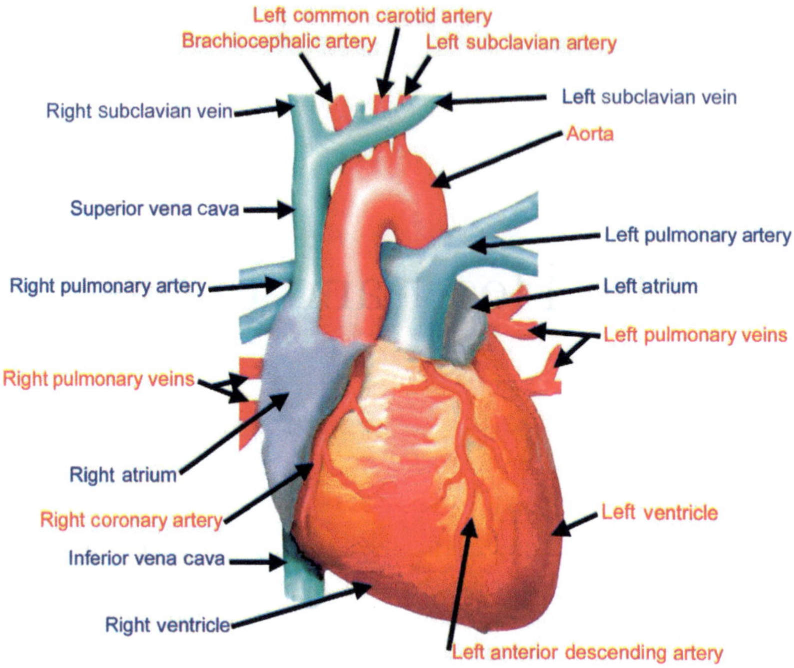

FIGURE 3.2.1 The heart.

blood due to blockage, the coronary arteries generally do not anastomose. Thus, blockage or myocardial infarction causes a heart attack.

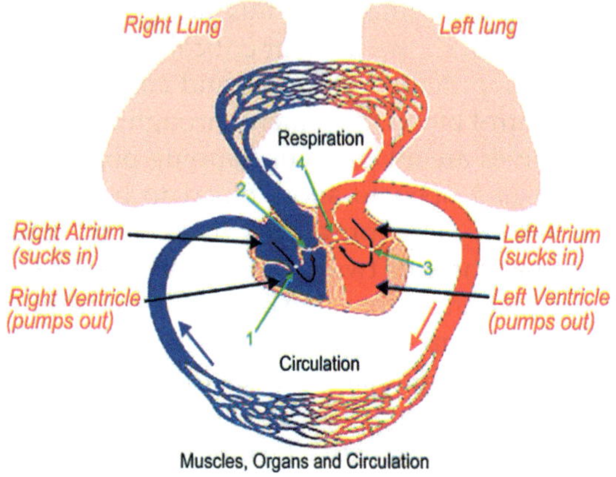

FIGURE 3.2.2 Figurative illustration of the inside of the heart and the circulation. The green arrows are 1. Tricuspid Valve, 2. Pulmonary Valve, 3. Mitral Valve, and 4. Aortic Valve.

In 2008, cardiovascular disease was the most common cause of death globally, accounting for 30% of fatalities. Three quarters of these fatalities were due to coronary artery disease or arterial blockages.

ARTERIES AND VEINS

As a reminder, an artery is any blood vessel taking blood from the heart, while a vein is any blood vessel bringing blood back to the heart. Figure 3.2.3 shows the major blood vessels in humans. Leaving the heart the main blood vessel carrying oxygenated blood is the aorta. Subdivisions of the aorta include the subclavian, common carotid, and femoral arteries (Figure 3.2.3). The pulmonary artery carries deoxygenated blood to the lungs. Returning to the heart the vena cava carries most of the body's blood. Subdivisions include the external and internal jugular veins,

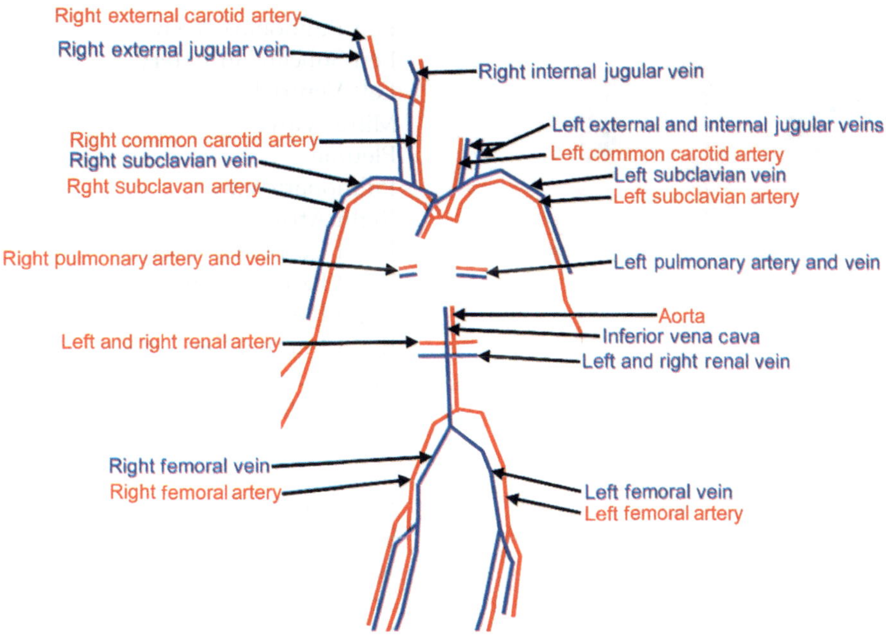

FIGURE 3.2.3 The major arteries and veins.

the subclavian and femoral veins. The pulmonary vein carries newly oxygenated blood from the lungs back to the heart (Figure 3.2.3).

THE LUNGS

The lung is the essential respiration organ in humans. The two lungs, left and right, are located near the backbone on either side of the heart. Their principal function is to transport oxygen from the atmosphere into the blood and to release carbon dioxide into the atmosphere. A large surface area is needed for this exchange of gases, which is accomplished by the specialized cells that form millions of tiny, exceptionally thin-walled air sacs called alveoli.

The air that enters through the mouth or the nose travels through the trachea, which divides into two the left and right lungs where they progressively subdivide into bronchi and bronchioles until the alveoli sacs are reached.

The lungs of humans have a soft, sponge-like texture and are honeycombed with epithelium, having a much larger surface area in total than the outer surface area of the lung itself. Breathing is largely driven by the muscular diaphragm at the bottom of the thorax. Contraction of the diaphragm pulls the bottom of the cavity in which the lung is enclosed downward, increasing volume and thus decreasing pressure, causing air to flow in. During normal breathing, expiration is passive and no muscles are contracted (the diaphragm relaxes).

Lung sacs are made up of clusters of alveoli, like grapes within a bunch. The individual alveoli are wrapped in blood vessels and it is here that gas exchange actually occurs. Deoxygenated blood from the heart is pumped through the pulmonary artery to the lungs, where oxygen diffuses into blood, attaches to hemoglobin,

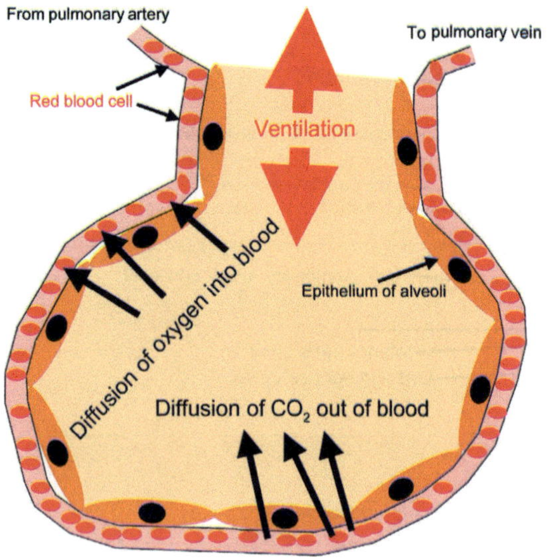

FIGURE 3.2.4 Gaseous exchange in alveolar sacs.

and is exchanged for blood born carbon dioxide (Figure 3.2.4). Each lung lobe is surrounded by a pleural cavity, which consists of two pleurae. The parietal pleura lies against the rib cage, and the visceral pleura lies on the surface of the lungs. In between the pleura is pleural fluid. The pleural cavity helps to lubricate the lungs, as well as provide the surface tension needed to keep the lung surface in contact with the rib cage.

CHAPTER KEYWORDS

Know these keywords and what each word refers to as well as its function. Read over the list. Are all these keywords familiar?

Alveoli
Aorta
Aortic Valve
Brachiocephalic Artery
Bronchi
Left Atrium
Left Common Carotid Artery
Left Pulmonary Artery
Left Pulmonary Vein
Left Subclavian Artery
Left Ventricle
Mitral Valve
Pleurae
Pulmonary Valve
Right Atrium
Right Coronary Artert
Right External carotid Artery
Right External Jugular Vein
Right Internal Jugular Vein
Right Pulmonary Artery
Right Pulmonary Vein
Right Renal Artery
Right Renal Vein
Right Ventricle
Sinoatrial node
Superior Vena Cava
Tricuspid Valve
Vagus Nerve

INTERNET REFERENCES

Use these references to find out more.
en.wikipedia.org/wiki/Heart
www.webmd.com/lung/picture-of-the-lungs
en.wikipedia.org/wiki/Lung
en.wikipedia.org/wiki/Heart_valve
en.wikipedia.org/wiki/Aorta
en.wikipedia.org/wiki/Subclavian_vein
en.wikipedia.org/wiki/Inferior_vena_cava

CHAPTER QUIZ

Answer these questions. Do you know the answers, holy cow! Check yourself out in this quiz, and prove that you are a humanology wiz.

1. The function of the right side of the heart is governed by which two valves?
a. Tricuspid and Pulmonary Valves
b. Mitral and Aortic valves

c. Tricuspid and Mitral Valves
d. Aortic and Pulmonary valves
e. Tricuspid and Aortic Valves

2. An incidental fast heart beat happens due to the
   a. Vagus nerve problem
   b. Brain signal problem
   c. Aorta problem
   d. Sino-atrial node problem
   e. Left ventricle of the heart problem
3. The Pulmonary Artery carries
   a. Oxygenated blood to the heart
   b. Oxygenated blood from the heart
   c. Deoygenated blood to the lungs
   d. Deoxygenated blood from the lungs
   e. Oxygenated blood from the heart to the head
4. The left internal jugular vein carries
   a. Deoxygenated blood from the left brain
   b. Oxygenated blood to the left brain
   c. Deoxygenated blood from the left arm
   d. Deoxygenated blood from the left thigh
   e. Oxygenated blood to the left thigh
5. The alveoli of the lungs are the site of
   a. Blood inter-exchanges with air oxygen
   b. Blood inter-exchanges with air CO_2
   c. Red blood cells being activated
   d. Oxygen and CO_2 inter-exchange between air and the blood
   e. Red blood cells being created

CHAPTER

3.3

The Circulation

THE BLOOD

Blood is a bodily fluid that delivers needed substances such as nutrients and oxygen to the cells and transports waste products away from those cells. Blood cells are suspended in plasma. Plasma constitutes 55% of blood volume; it is mostly water (92% by volume), and contains proteins, glucose, minerals, hormones, and carbon dioxide. Albumin is the main protein in blood and functions to regulate the colloidal osmotic neutrality of blood. Blood cells are mainly red blood cells, correctly called erythrocytes, and white blood cells including leukocytes and platelets. Erythrocytes contain hemoglobin, an iron-containing protein, which facilitates transportation of oxygen, a gas that otherwise would not be soluble in blood. In contrast, carbon dioxide is almost entirely transported outside of blood cells, dissolved in liquid plasma as complex of the bicarbonate ion ($CO_2 + H_2O = H^+ + HCO_3^-$). Humans have an adaptive immune system based largely on white blood cells. White blood cells help to resist infections. White blood cell platelets are an essential part of clotting of blood.

Blood performs many essential functions within the body including supply of oxygen to tissues; supply of nutrients to the body; removal of waste such as carbon dioxide, urea, and lactic acid from the body; immunological functions including detection of foreign material by antibodies; phagocytosis, a major mechanism used to remove pathogens and cell debris; coagulation, the response to a broken blood vessel; and messenger functions such as the transport of endocrines throughout the body.

ERYTHROCYTES

Erythrocytes are flying saucer or discoid structures about 0.007 mm in diameter and 0.001 mm thick. The erythrocyte is like a big sack. It contains no nucleus and no organelles (no DNA) and is stuffed with hemoglobin molecules. Humans have $20\text{-}30 \times 10^{12}$ erythrocytes in their blood. The total surface area on erythrocytes is proportional to 2000 times the body's total exterior surface, illustrating the efficiency of hemoglobin at carrying oxygen. It is estimated that each erythrocyte takes approximately 20 s to circulate the blood and pass through the heart and lungs.

With no nucleus, DNA, or organelles, erythrocytes cannot duplicate themselves. Instead, they

Human Physiology, Biochemistry and Basic Medicine
http://dx.doi.org/10.1016/B978-0-12-803699-0.00026-8

are produced by a process called erythropoiris. They are made in the red bone marrow of large bones and take approximately 7 days to be created. Red bone marrow makes approximately two million new erythrocytes per second or about 7 billion new erythrocytes per hour.

Human blood is red because of the erythrocyte content, which contains hemoglobin. Hemoglobin is a red-colored iron-containing protein. Oxygenated hemoglobin is bright red. Deoxygenated hemoglobin is not blue as once thought, it is dark red. Human blood does not appear blue under any circumstances. Deoxygenated blood contains roughly 32% of oxygen and freshly oxygenated blood contains 98% oxygen. Both are shades of red. While your blood never actually turns blue, your skin can take on a bluish cast as a result of certain diseases and disorders. This blue color is called cyanosis.

Hemoglobin is a large protein of molecular weight 64,000. It comprises four peptide strings, two α-subunits and two β-subunits, containing 574 amino acids total. Each of the four subunits contains a heme component. Figure 3.3.1 shows a three-dimensional model of the structure of hemoglobin. Each of the four subunits contains one heme molecule, a complex carbon, and nitrogen complex anchoring one ferrous or iron molecule (Figure 3.3.2). Each iron molecule

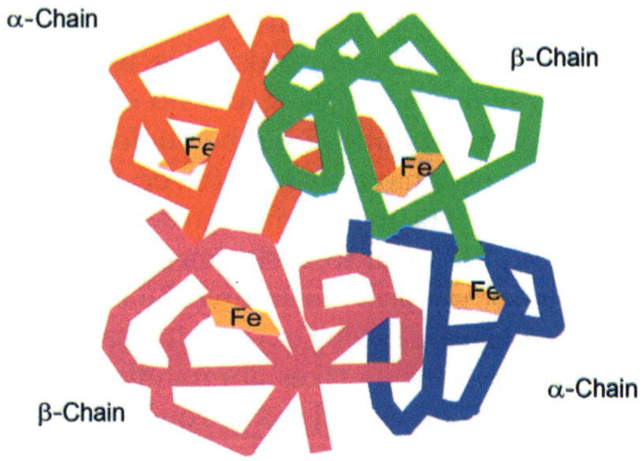

FIGURE 3.3.1 Three-dimensional model of the large protein hemoglobin.

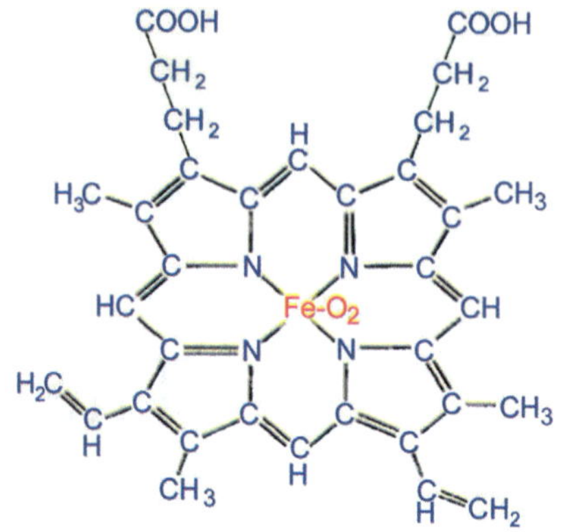

FIGURE 3.3.2 Heme components of hemoglobin.

holds one diatomic oxygen molecule (O_2) as illustrated in Figure 3.3.2. Therefore, each hemoglobin molecule can hold four molecules of O_2. With this design, erythrocytes very efficiently carry O_2. Very few gases other than oxygen bind the heme of hemoglobin. Exceptions include carbon monoxide and cyanide, which are highly poisonous and most people stay very clear from. Carbon dioxide is not transported by erythrocytes; it is dissolved in blood as a complex of the bicarbonate ion.

Human erythrocytes are produced through a process called erythropoiesis, developing from stem cells to mature erythrocytes in about 7 days. When matured, these cells live in blood circulation for about 80-120 days. At the end of their life span, they become senescent and are removed from circulation.

In the process of erythropoiesis, a red bone marrow cell undergoes a series of differentiations (Figure 3.3.3). A bone marrow Hemocytoblast is the stem cell. This is differentiated into a multipotent stem cell, then a unipotent stem cell, and then a proerythroblast. This becomes an early erythroblast, then a late normoblast at which stage hemoglobin accumulates in the cell. Then the remaining nucleus is expelled before the cell becomes a reticulocyte (Figure 3.3.3).

The reticulocyte is released from the bone marrow. After 1-2 days, this becomes

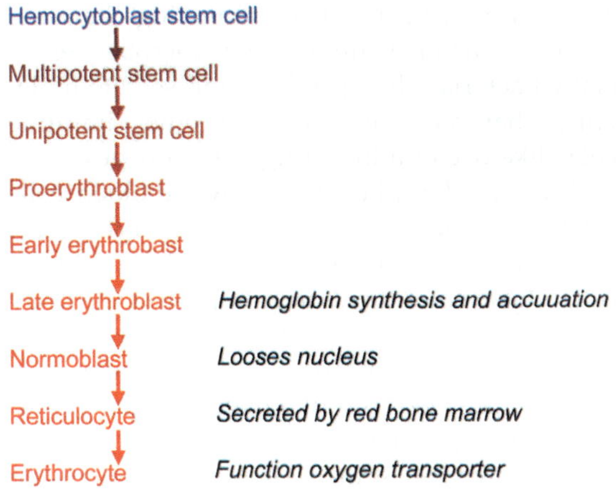

FIGURE 3.3.3 Erythropoiesis and erythrocyte formation in bone marrow.

erythrocytes or mature red blood cells (Figure 3.3.3). Essential for the maturation of erythrocytes are Vitamin B12 (cobalamin) and Vitamin B9 (Folic acid).

THE IMMUNE SYSTEM

The purpose of the immune system is to avert virus, bacterial, fungi, or other infections. The immune system can distinguish between normal body cells and unhealthy body cells by recognizing danger signals called Danger-Associated Molecular Patterns (DAMPs). Cells may be unhealthy because of infection or because of cellular damage caused by other agents like sunburn or cancer. Infectious microbes like viruses and bacteria release a separate set of signals recognized by the immune system called Pathogen-Associated Molecular Patterns (PAMPs).

As the immune system detects these signals, it reacts to the problem. If an immune response cannot be activated, problems arise, such as infection. In contrast, when an immune response is activated without a real threat different problems arise, such as allergic reactions and autoimmune disease. There are numerous cell types in the immune system. Each cell type plays a unique role with different ways of recognizing problems and performing functions.

All immune system cells derive from precursor cells in bone marrow and develop into mature cells through a series of changes that occur in different parts of the body. The bone marrow contains stem cells that can develop into a variety of cell types. The myeloid progenitor stem cell in the bone marrow is the precursor to innate immune cells, to neutrophils, eosinophils, basophils, mast cells, monocytes, dendritic cells, and macrophages, which are important first-line responses to infection.

The lymphoid progenitor stem cell leads to adaptive immune cells, to B-cells and T-cells that are responsible for mounting responses to microbes based on immunological memory. Natural killer cells also are derived from the common lymphoid progenitor and share features of both innate and adaptive immune cells, providing immediate defenses like innate cells but also may be retained as memory cells like adaptive cells. B-, T-, and natural killer cells also are called lymphocytes.

Immune cells circulate through the bloodstream constantly, searching for problems. T cells mature in the thymus, a small organ located in the upper chest. The lymphatic system is a network of vessels and tissues composed of lymph, an extracellular fluid, and lymphoid organs, such as lymph nodes. The lymphatic system is a conduit for communication between tissues and the bloodstream. Immune cells are carried through the lymphatic system and converge in lymph nodes, which are found throughout the body.

Lymph nodes are a communication hub where immune cells sample information brought in from the body. For instance, if adaptive immune cells in the lymph node recognize pieces of a microbe brought in from a distant area, they will activate, replicate, and leave the lymph node to circulate and address the pathogen. Thus, physicians may check patients for swollen lymph nodes, which can indicate an active immune response.

Reflecting on all the different white blood cells that manage the immune system.

Granulocytes, including *Basophils*, *Eosinophils*, and *Neutrophils*, are important for host defense against parasites. Granulocytes can phagocytose, or ingest, bacteria, degrading them. Neutrophils accumulate at sites of local tissue injury and then communicate with each other to form cellular swarms.

Mast cells also are important for defense against all parasites. Mast cells mediate allergic reactions by releasing inflammatory chemicals like histamine.

Monocytes develop into *Macrophages* and also patrol and respond to all problems. Macrophages ingest and degrade bacteria. Macrophages also have important nonimmune functions, such as recycling dead cells and clearing away cellular debris.

Dendritic cells are an important antigen-presenting cell. Antigens are molecules from pathogens, host cells, and allergens that may be recognized by adaptive immune cells.

Natural killer cells are important for identifying virus-infected cells. They contain intracellular compartments called granules, which are filled with proteins that can make holes in the target cells and also cause apoptosis.

B-cells have two major functions. They present antigens to T-cells, and more importantly, they produce antibodies to neutralize infectious microbes. Neutralization occurs when the pathogen is covered in antibodies. Antibodies are expressed in two ways. The B-cell receptor (BCR), which sits on the surface of a B-cell, is an antibody. B-cells also secrete antibodies to bind to pathogens.

T-cells have a variety of roles and are classified by subsets. T-cells are divided into two broad categories, CD8+ T-cells and CD4+ T-cells, based on which protein is present on the cell's surface. CD8+ T-cells also are called cytotoxic T-cells. They are crucial for recognizing and removing virus-infected cells. The CD4+ T-cell are critical for coordinating immune responses against intracellular microbes, especially bacteria. They produce and secrete molecules that alert and activate other immune cells, like bacteria-ingesting macrophages.

The term lymphocytes for white blood cells refers to natural killer cells, B-cells, and T cells. The term leukocyte refers to a colorless blood cell, whose chief function is to protect the body against microorganisms causing disease, or to all cells of the immune system. Leukocytes may be classified into two main groups: granular and nongranular.

As shown in Figure 3.3.4, multiple organs tie together the human immune system or human lymphoid system. The thymus is the suite of formation of T-cells. The spleen filters the blood and distributes T- and B-cells. The spleen protects the body against bacterial infection. Lymph nodes are the site of storage and formation of lymphocytes, natural killer cells, T-cells, and B-cells. The bone marrow is the site of B-cell production. Lymphoid tissue is found in other sites, Peyer's patch, the Appendix, and the Tonsils and Adenoids (Figure 3.3.4). Lymph nodes are found in the neck, under the armpits, and adjacent to the groin.

COAGULATION

Coagulation is the method by which blood changes from a liquid to a dense gel. It results in stop of blood loss, hemostasis, from a damaged vessel, and is followed by repair. The pathway of coagulation involves activation, adhesion, and aggregation of platelets along with depositing and maturation of the fibrous protein fibrin.

Coagulation starts instantly after a blood vessel injury. Exposure of blood to the space under the endothelium initiates two processes. First, changes in platelets, and the exposure of subendothilial tissue factor to plasma Factor 7, leads to fibrin formation. Platelets immediately form a plug at the site of injury. Additional coagulation

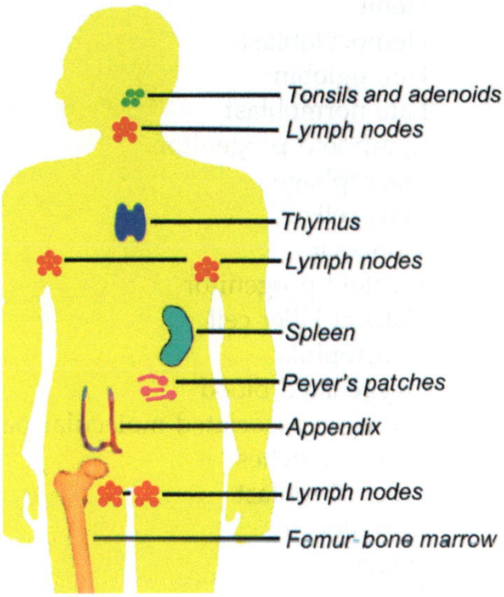

Thymus Site of formation of T cells.

Spleen Filters blood and distributes T and B cells. Protects body against bacterial infection.

Lymph nodes Lymph nodes become enflamed and enlarged in response to foreign tissue such as cancer, and to infection. This is the site of storage and formation of lymphocytes, natural killer cells, T and B cells.

Bone marrow Site of B cell production.

Peyer's patches Organized lymphoid nodules or aggregated lymphoid tissue. Located next to intestines.

Appendix The gut lymphoid tissue.

Tonsils and adenoids Lymphoid tissue posterior to nasal cavity. The immune systems first line of defense against ingested pathogens.

FIGURE 3.3.4 The organs of the immune system.

factors or clotting factors other than Factor 7 respond in a cascade to form fibrin strands, which strengthen the platelet plug. When blood vessels are cut or damaged, the loss of blood from the system must be stopped before shock and possible death occur. This is accomplished by solidification of the blood, a process called coagulation or clotting.

A blood clot comprises a plug of platelets enmeshed in a network of insoluble fibrin molecules. Platelet aggregation and fibrin formation both require the protease or degradative enzyme thrombin. Clotting also requires a variety of clotting factors to form the fibrin and the clot (Figure 3.3.5). These circulate in the blood as inactive precursors. They are activated by proteases becoming, in turn, active proteases for other factors in the system (Figure 3.3.5).

There are two processes that can initiate clotting: a rapid process called the extrinsic pathway and a slower process called the intrinsic pathway. In the extrinsic pathway, damaged cells display a tissue factor that activates the coagulation process by activating Factor 7. Factor 7 is a protease that can act in two ways. It either acts directly on Factor 10 or acts on Factor 9, which in turn acts on Factor 10 (Figure 3.3.5). Factor 10 binds and activates Factor 5, which forms Thrombin. Thrombin has several different activities. Two of them are proteolytic cleavage of Fibrinogen to form Fibrin and platelet activation (Figure 3.3.5). From the activated platelets and the fibrin the clot is formed.

In the intrinsic or slower pathway, Factor 12 in blood escapes into injured tissue spaces and contacts with collagens. Activated Factor 12 is a protease that activates Factor 11 which, in turn, activates Factor 9, then Factor 10 and proceeds down the same activation route as the extrinsic pathway (Figure 3.3.5).

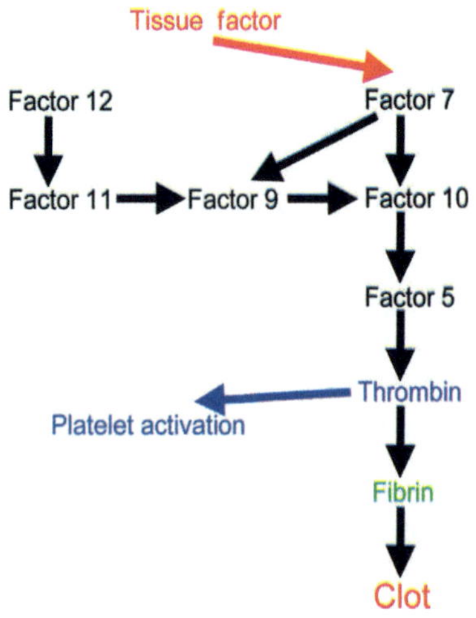

FIGURE 3.3.5 Blood coagulation factors in the clotting process.

CHAPTER KEYWORDS

Know these keywords and what each word refers to as well as its function. Read over the list. Are all these keywords familiar?

B-cell
Basophils
CD4+ T cell
CD8+ T cell
DAMP
Danger-associated molecular pattern
Dendritic cell
Deoxygenated blood
Eosinophils
Erythrocyte
Erythropoiesis
Factor 10
Factor 11
Factor 12
Factor 5
Factor 7
Factor 9
Fibrin
Granulocytes
Heme
Hemocytoblast
Hemoglobin
Late normoblast
Lymphoid progenitor
Macrophage
Mast cell
Monocyte
Myeloid progenitor
Natural killer cell
Neutrophils
Oxygenated blood
Pathogen-associated molecular pattern
Peyer's patches
Proerythroblast
Reticulocyte
Spleen
T-cell
Thrombin
Tissue factor
Unipotent stem cell
Vitamin B12
Cobalamin
Vitamin B9
Folic acid
α-Chain
β-Chain

INTERNET REFERENCES

Use these references to find out more.
www.medicinenet.com/hemoglobin/article.htm
en.wikipedia.org/wiki/Hemoglobin
en.wikipedia.org/wiki/Red_blood_cell
www.medicinenet.com/script/main/art.asp?articlekey=3309
en.wikipedia.org/wiki/Immune_system
www.niaid.nih.gov/topics/immunesystem/Pages/default.aspx
en.wikipedia.org/wiki/Coagulation
en.wikipedia.org/wiki/Erythropoiesis

CHAPTER QUIZ

Answer these questions. Do you know the answers, holy cow! Check yourself out in this quiz, and prove that you are a humanology wiz.

1. Which statement is false regarding erythrocytes?
   - **a.** They are red blood cells.
   - **b.** They contain no nucleus.
   - **c.** They are stuffed full of hemoglobin.
   - **d.** They contain no mitochondria.
   - **e.** The have a membrane made of iron.
2. Which statement is false regarding hemoglobin?
   - **a.** It is a large four subunit protein.
   - **b.** It contains four heme elements.
   - **c.** It is a large two subunit protein.
   - **d.** It contains the element iron.
   - **e.** It contains iron that binds oxygen.
3. Which statement is true about natural killer cells?
   - **a.** They contain granules that can make holes in the target cell.
   - **b.** They generate antibodies to the target cells.
   - **c.** They poison and kill target cells using granulopin.
   - **d.** They digest derade and destroy parasite cells.
   - **e.** They destroy the plasma membrane of target cells using a protease.
4. Which statement is true about macrophages?
   - **a.** They contain granules that can make holes in the target cell.
   - **b.** They generate antibodies to the target cells.
   - **c.** They poison and killer target cells using granulopin.
   - **d.** They digest degrade and destroy parasite cells.
   - **e.** They destroy the plasma membrane of target cells using a protease.
5. In the coagulation process which factor initiates rapid clot formation?
   - **a.** Factor 11
   - **b.** Factor 1
   - **c.** Factor 7
   - **d.** Factor 4
   - **e.** Factor 5

CHAPTER

3.4

Brain and Nervous System

THE NERVOUS SYSTEM

A neuron or nerve cell is an electric transmitting cell that processes information in the nervous system through electrical and chemical ion signals. The electrical signals between neurons occur via synapses, specialized connections with other neuron cells. Neurons are the main components of the nervous system, which includes the brain, spinal cord, and the peripheral nervous system, which comprises the central nervous system.

As shown in Figure 3.4.1, a neuron comprises a head and its feelers or dendrites and a long axon or wire protected by a myelin sheath. The tail comprises dendrites or terminal ending buttons. The synapse or the nerve-nerve connection is a structure that permits a neuron to pass an electrical or chemical signal to another cell. Synapses are essential to neuronal function; they pass signals to individual target cells. At a synapse, the plasma membrane of the signal-passing neuron comes into close contact with the membrane of the target cell. Both cells contain extensive arrays of molecular machinery that link the two membranes together and carry out the signaling process. Synapses transmit chemical signal using neurotransmitters. The neurotransmitter released by the pre-synaptic neuron crosses the synapse and acts on a receptor on the post-synaptic neuron. Common nerve-nerve neurotransmitters include the amino acids Glu and Asp, the catecholamines epinephrine, norepinephrine and dopamine, and acetylcholine.

Specialized types of neurons include: sensory neurons that respond to touch, sound, light, and other stimuli; motor neurons that receive signals from the brain and spinal cord and cause muscle contractions; and interneurons that connect neurons to other neurons within the same region of the brain or spinal cord in neural networks. Individual neurons may be up to a meter long. When one considers the millions of neurons that interconnect brain memories and brain thoughts, each human may be wired with 100,000 miles, four times around the planet earth, of neurons.

All neurons are electrically excitable, maintaining voltage gradients across their membranes by means of ion pumps, such as sodium, potassium, chloride, and calcium ions. Neurons interconnect brain memories and ideas and muscles in the body. Neurons start in the brain and head down the spinal cord and then generally follow arteries and veins, supplying nervous signals and nervous detectors to the extremities of the body.

Human Physiology, Biochemistry and Basic Medicine
http://dx.doi.org/10.1016/B978-0-12-803699-0.00027-X

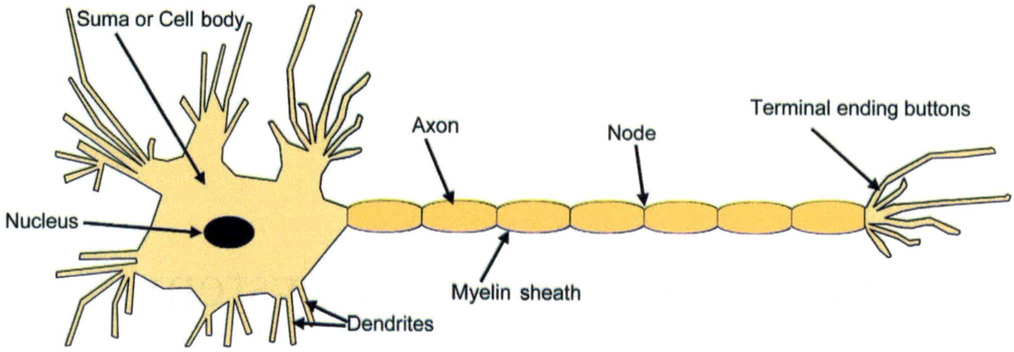

FIGURE 3.4.1 A neuron or nerve cell.

THE BRAIN

The brain is by far the most complex organ in a human body. While the brain appears anatomically as a solid mass, it is actually a mix of hundreds of billions of neurons and synapses, largely interconnected in different ways. It is like a solid mess of wires. We divide the brain into the frontal lobe, parietal lobe, occipital lobe, cerebellum, temporal lobe, midbrain, pons, medulla, thalamus, fornix, mammillary body, cingulate gyrus, hypothalamus, and hippocampus for means of analysis. These divisions are illustrated in Figures 3.4.2 and 3.4.3.

The *Hypothalamus* is located in the middle of the brain at the bottom (Figure 3.4.3). In humans, it is roughly the size of an almond. The hypothalamus is a portion of the brain that contains a number of small nuclei with many functions. One of the functions of the hypothalamus is to link the nervous system to the endocrine system via the pituitary gland. The hypothalamus is responsible for multiple activities of the autonomic nervous system. It synthesizes and secretes certain neurohormones, and these in turn stimulate or inhibit the secretion of pituitary hormones. While the hypothalamus is neuron brain tissue is secretes four hormone like endocrine tissue. The hypothalamus controls body temperature, hunger, thirst, fatigue, sleep, and circadian rhythms.

The anterior side of the hypothalamus includes the preoptic nucleus. This has testosterone receptors and through testosterone action controls sexual libido. Nearby regions, perhaps a millimeter away, are the medial preoptic nucleus. This produces the hypothalamic-releasing

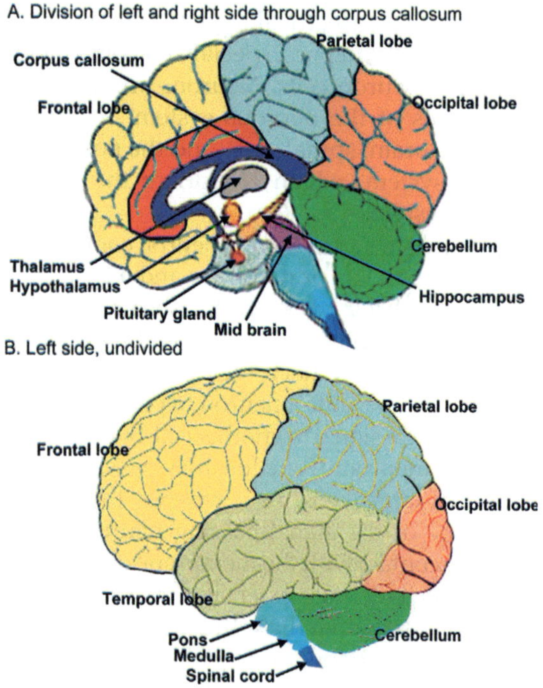

FIGURE 3.4.2 Anatomy of the brain.

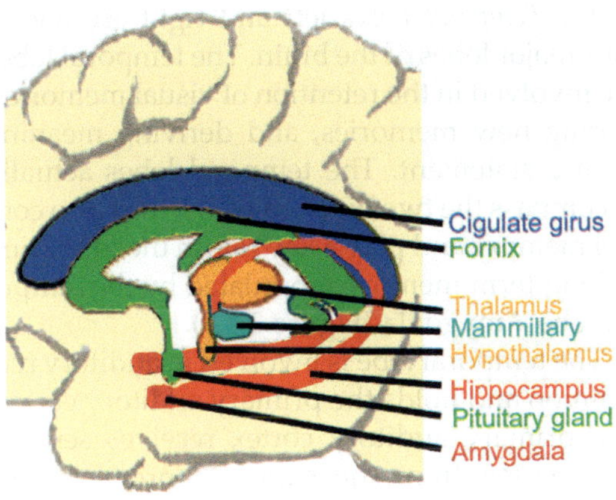

FIGURE 3.4.3 The anatomy of the core of the center of the brain.

hormone, gonadotropin-releasing hormone (GnRH), which controls the reproductive system in men and women. Nearby, the paraventricular nucleus produces other releasing hormones: thyrotropin-releasing hormone (TRH), corticotropin-releasing hormone (CRH), and prolactin-releasing hormone (PRLH). These releasing hormones are secreted into the hypothalamic-hypophyseal portal circulation and act on the anterior pituitary.

In the anterior hypothalamus is the anterior hypothalamus nucleus, which controls panting and sweating in humans. This nucleus controls the body temperature and its constancy. Temperature is controlled through release and action of prostaglandin E2. In many ways the hypothalamus functions like a thermostat. When the set point or hypothalamic thermostat is raised, the body increases its temperature through activity generation of heat and retaining heat. Vasoconstriction both reduces heat loss through the skin and causes the person to feel cold. In the anterior lateral region of this almond size structure, the hypothalamus, the lateral nucleus controls human thirst and human hunger responses.

On the tuberal side of the hypothalamus, the dorsomedial hypothalamic nucleus has a neurological control over one's heart beat and blood pressure. This arcuate nucleus controls release of growth hormone releasing hormone (GHRH) through the hypothalamic-hypophyseal portal circulation. The lateral nucleus on the tuberal side of the organ is also involved in control of thirst and hunger. On the posterior side of the organ, the posterior nucleus controls bodily shivering and eye dilation.

The *Hippocampus*, located just to the side of the hypothalamus, is another critical area of the brain (Figure 3.4.3). Humans have a left and right hippocampus, one in each side of the brain, which are two coil- or curve-shaped structures. They play important roles in the consolidation of information from short-term memory to long-term memory or memory management. People with extensive hippocampal damage may experience amnesia or the inability to form or retain new memories.

Psychologists and neuroscientists generally agree that the hippocampus plays an important role in the formation of all new memories. Part of this function is hippocampal involvement in the detection of novel events, places, and stimuli. Some researchers regard the hippocampus as part of a larger temporal lobe memory system. It is responsible for controlling all memory. Some researchers distinguish between conscious recollection, which depends on the hippocampus, and familiarity, which depends on portions of the temporal lobe.

Spatial firing patterns of eight place cells have been recorded using rats. The rat ran back and forth along an elevated track, stopping at each end to eat a small food reward. Studies conducted on freely moving rats and mice have shown that many hippocampal neurons have "place fields," i.e., they fire bursts of action potentials when a rat passes through a particular part of the environment.

The discovery of place cells in the 1970s led to a theory that the hippocampus might also act as a cognitive map—a neural representation of the layout of the environment. Several lines of evidence support this hypothesis. It is a frequent observation that without a fully functional hippocampus, humans may not remember where they have been and how to get where they are going. Brain imaging shows that people have more active hippocampi when correctly navigating, as tested in a computer-simulated "virtual" navigation task. Also, there is evidence that the hippocampus plays a role in finding shortcuts and new routes between familiar places.

The *Thalamus* is located over the hypothalamus and exists as two halves (Figure 3.4.3). The thalamus has multiple functions; it is like a kind of brain switchboard. It is generally believed to act as a relay between different lobes of the brain. Considering the visual system, for example, inputs from the retina are sent to the lateral geniculate nucleus of the thalamus, which in turn projects them to the occipital lobe. The thalamus is believed to both process sensory information as well as relay it.

The thalamus also plays an important role in regulating states of sleep and wakefulness. Thalamic nuclei have strong reciprocal connections with the lobes of the brain, forming circuits that are believed to be involved with consciousness. The thalamus plays a major role in regulating arousal, the level of awareness, and activity. Damage to the thalamus leads to coma. It is thought that the coordination of the brain and the concepts of consciousness lie in the thalamus.

The *Mammillary Body* is a projection from the thalamus that is important for recollective memory, the where and when of memories, and spatial memory, the perception of environment and spatial orientation (Figure 3.4.3).

The *Fornix* is a bundle of neurons that synchronizes and signals the hippocampus and mammillary body. This coordinates and links the memory system (Figure 3.4.3).

The *Cigulate Gyrus* is the part of the brain that deals with pain and pain sensation and manages emotions associated with pain (Figure 3.4.3).

The Temporal Lobes left and right are one of four major lobes of the brain. The temporal lobes are involved in the retention of visual memories, storing new memories, and deriving meaning from a statement. The temporal lobes actually encompass the two hippocampi, which also control memory and play a key role in the formation of long-term memory modulated by the temporal lobe Amygdala (Figure 3.4.3).

The temporal lobe is involved in auditory recognition and holds the primary auditory cortex. The primary auditory cortex receives sensory information from the ears and processes the information into meaningful units such as speech and words. The temporal lobes include an area where auditory signals from the ear first reach the brain and are processed by the primary auditory cortex.

The areas of the temporal lobes associated with vision interpret the meaning of visual stimuli. The temporal lobes appear to be involved in high-level visual processing of complex stimuli such as faces. Anterior parts of the temporal lobes are used for visual processing and are involved in object recognition.

The medial temporal lobes are thought to be involved in long-term memory. The medial temporal lobes include the hippocampi, which are essential for memory storage. Therefore, damage to this area can result in impairment in memory, leading to permanent anterograde amnesia.

The *Cerebellum* is at the rear of the brain. It is an important region involved in motor control of the body. Clues regarding the function of the cerebellum have come from the consequences of damage to it. Humans with cerebellar dysfunction show, above all, problems with motor control, on the corresponding side of the body. They lose precision, producing erratic, uncoordinated, or incorrectly timed movements.

Functional imaging studies have shown cerebellar activation in relation to language, attention, and mental imagery; correlation studies have shown interactions between the cerebellum and non-motoric areas of the cerebral cortex; and a variety of non-motor symptoms have been recognized in people with damage that appears to be confined to the cerebellum.

The *Medulla*, or the medulla oblongata, principally connects the higher levels of the brain to the spinal cord, and is responsible for several functions of the autonomic nervous system. These include the respiration chemoreceptors. These chemoreceptors detect changes in the acidity of the blood. If the blood is considered too acidic by the medulla, electrical signals are sent to the lungs increasing their contraction rate in order to reoxygenate the blood.

The *Pons* is a region of the brain at the top of the spinal cord. The pons contains nuclei that relay signals from the forebrain to the cerebellum. The pons is the root of four cranial nerves managing hearing, equilibrium, and taste, and in facial sensations such as touch and pain, as well as motor roles in eye movement, facial expressions, chewing, swallowing, and the secretion of saliva and tears.

The *Frontal lobe*, *Occipital lobe*, and *Parietal lobe* are all parts of the Cerebral Hemispheres or Cerebral cortex of the brain. These are places where long- and short-term memories are stored. The left and right hemispheres or cerebral cortex of the brain are linked together through the corpus callosum. The Cerebral hemispheres or Cerebral cortex are places where simple calculations are made and where short-term and long-term memory is stored. Memory is processed through the hippocampi and the temporal lobe processing centers and can be classified as short-term memory and long-term memory.

Short-term memory allows recollection for a period of seconds or a minute without rehearsal. Short-term memory relies mostly on an acoustic code for storing information, and to a lesser extent a visual code. Short-term memory generally has a limited storage capacity and duration, indicating that information is not retained indefinitely. By contrast, long-term memory can store much larger amounts of information for possibly unlimited duration. The capacity of long-term memory is immeasurably large. For example, we may remember a random seven-digit number for only a few seconds before forgetting it, suggesting it was stored in our short-term memory. In contrast, we can remember telephone numbers for many years through repetition; this information is said to be stored in long-term memory. All of this is stored in the Cerebral hemispheres or Cerebral cortex.

Short-term memory is supported by patterns of neuronal communication, dependent mostly on regions of the frontal lobe of the brain. Long-term memory, on the other hand, is maintained by more stable and permanent changes in neural connections widely spread throughout the brain. The hippocampi are essential to the consolidation of information from short-term to long-term memory, but do not store information therein. Without the hippocampi, new memories cannot be stored into long-term memory. One of the main functions of sleep is the improvement of the consolidation of information, as several studies have demonstrated that memory depends on having sufficient sleep between training and testing.

Research suggests that during sleep long-term memory may be programmed. According to the Multi-trace distribution model, all memories of a single event may be stored in multiple close-by locations. For instance, as illustrated in Figure 3.4.4, a bike-riding experience may be stored in the Cerebral hemispheres and the memory may be stored in multiple inter-connected locations. How to ride a bike may be stored in one location, how to travel between two locations may be stored elsewhere, biking safety issues may be stored in another location, the emotion of biking in another location, sights seen on the biking trip in a further location, and sound heard on the biking trip in a final location. All of the events are inter-connected by neurons.

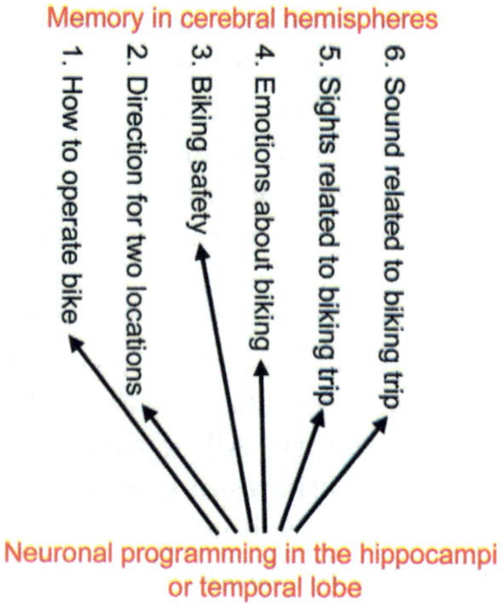

FIGURE 3.4.4 Multi-trace distribution model, example memories of biking trip.

CHAPTER KEYWORDS

Know these keywords and what each word refers to as well as its function. Read over the list. Are all these keywords familiar?

Amygdala
Anterior hypothalamus nucleus
Autonomic nervous system
Axon
Cerebellum
Corticotropin-releasing hormone
Cigulate Gyrus
Dendrite
Dorsomedial hypothalamic nucleus
Fornix
Frontal lobe
Gonadotropin-releasing hormone
Growth hormone-releasing hormone
Hippocampus
Hypothalamus
Lateral nucleus
Medial preoptic nucleus
Medulla
Medulla oblongata
Midbrain
Multi-trace distribution model
Myelin sheath
Neurohormone
Neuron
Neurotransmitter
Occipital lobe
Parietal lobe
Pons
Preoptic nucleus
Suma
Synapse
Temporal lobe
Terminal ending button
Thalamus
Thyrotropin-releasing hormone

INTERNET REFERENCES

Use these references to find out more.
en.wikipedia.org/wiki/Hypothalamus
en.wikipedia.org/wiki/Hippocampus
en.wikipedia.org/wiki/Thalamus
en.wikipedia.org/wiki/Cerebellum
en.wikipedia.org/wiki/Frontal_lobe
en.wikipedia.org/wiki/Parietal_lobe
en.wikipedia.org/wiki/Occipital_lobe
en.wikipedia.org/wiki/Temporal_lobe
en.wikipedia.org/wiki/Memory
en.wikipedia.org/wiki/Brain
en.wikipedia.org/wiki/Brain_stem
en.wikipedia.org/wiki/Cerebral_cortex

CHAPTER QUIZ

Answer these questions. Do you know the answers, holy cow! Check yourself out in this quiz, and prove that you are a humanology wiz.

1. The hypothalamus has what function?
   a. It is a motor control region of the brain.
   b. It controls sexual libido.
   c. It controls long-term memory.
   d. It controls four cranial nerves.
   e. It relays information between different regions of the cerebral cortex.

2. The hippocampi have what function?
  a. It is a motor control region of the brain.
  b. It controls sexual libido.
  c. It controls long-term memory.
  d. It controls four cranial nerves.
  e. It relays information between different regions of the cerebral cortex.

3. The thalamus has what function?
  a. It is a motor control region of the brain.
  b. It controls sexual libido.
  c. It controls long-term memory.
  d. It controls four cranial nerves.
  e. It relays information between different regions of the cerebral cortex.

4. The cerebellum has what function?
  a. It is a motor control region of the brain.
  b. It controls sexual libido.
  c. It controls long-term memory.
  d. It controls four cranial nerves.
  e. It relays information between different regions of the cerebral cortex.

5. The corpus callosum
  a. Is part of the temporal lobe
  b. Is part of the brain stem
  c. Links the left and right hemispheres of the brain
  d. Control the short-term memory
  e. Controls thirst and hunger

CHAPTER

3.5

Seeing, Hearing, Tasting, Smelling, and Touching

SEEING

Eyes detect light and convert it into electrochemical impulses in neurons. In humans, the eye is a complex optical system that collects light from the surrounding environment, regulates its intensity through a diaphragm, focuses it through lenses to form an image, converts this image into a set of electrical signals, and transmits these signals to the brain through complex neural pathways that connect the eye via the optic nerve to the brain.

The human eye works much like a modern digital camera (Figure 3.5.1). Light is focused by the cornea, the clear front surface of the eye, which acts like a camera lens. The iris of the eye functions like the diaphragm of a camera, controlling the amount of light reaching the back of the eye by automatically adjusting the aperture of the pupil. The eye's crystalline lens is located directly behind the pupil and further focuses light; this lens helps the eye automatically focus on near and approaching objects. Light is focused on the retina, the light-sensitive inner lining of the back of the eye (Figure 3.5.1). The retina acts like an electronic image sensor converting it into electronic signals. The optic nerve then transmits these signals to the brain.

The retina contains two major types of light-sensitive cells used for vision, the rods and the cones. Rods cannot distinguish colors, but are responsible for black-and-white vision. Rods work well in dim light because they contain the pigment rhodopsin, which is sensitive at low light intensity. Rods are distributed throughout the retina, but there are none at the fovea and none at the blind spot.

Cones control color vision in humans. They require brighter light to operate than rods. In humans, there are three types of cones: maximally sensitive to long-wavelength, medium-wavelength, and short-wavelength light, or red, green, and blue, colors. Objects are seen most sharply in focus when their focused images fall on the fovea. Cone cells and rods are connected through intermediate cells in the retina to nerve fibers of the optic nerve.

Human Physiology, Biochemistry and Basic Medicine
http://dx.doi.org/10.1016/B978-0-12-803699-0.00028-1

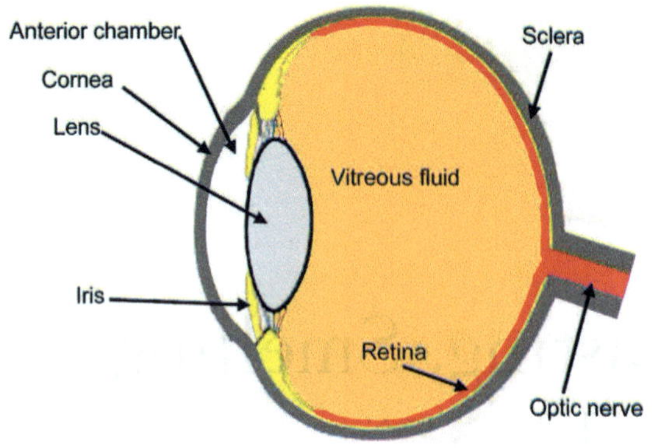

FIGURE 3.5.1 Anatomy of the human eye.

HEARING

The anatomy of the ear is shown in Figure 3.5.2. Auditory vibrations resonate against the tympanic membrane. The sounds are then resonated through the three ear bones, the malleus, incus, and stapes, in the middle ear. The malleus is linked to the tympanic membranes and the stapes is linked to the cochlea. The sound then resonates into the cochlea or inner ear where it is detected by the cochlear nerve and transmitted to the brain (Figure 3.5.2).

The semicircular canals of the cochlea or inner ear cope with the senses of balance and motion along with the vestibular nerve.

TASTING

Taste perception or gustation is the sensory detection of food on the tongue. Taste is the sensation that occurs in the mouth when a substance reacts chemically with taste receptor cells located on taste buds or papillae (Figure 3.5.3). Taste determines flavors of foods. Humans have taste receptors on taste buds in the upper surface of the tongue and the epiglottis.

The tongue is covered with thousands of small bumps called papillae, which can be seen by the naked eye (Figure 3.5.3). These taste buds or papilla are anchored to the afferent nerve to signal the brain. There are between 2000 and 5000 taste buds on the back and front of the tongue.

The sensation of taste can be categorized as sweetness, sourness, saltiness, bitterness, and umami (Glu-related Asian food taste). Sweet, savory, and bitter tastes are triggered by binding of molecules to G protein-coupled receptors on the cell membranes of taste buds. Saltiness and sourness are perceived when alkali metal or hydrogen ions enter the taste buds. The basic tastes are only part of the sensation and flavor of food in the mouth. Temperature may be detected by thermoreceptors in the mouth. Taste perception fades with age: on average, people lose half their taste receptors by the time they turn 20.

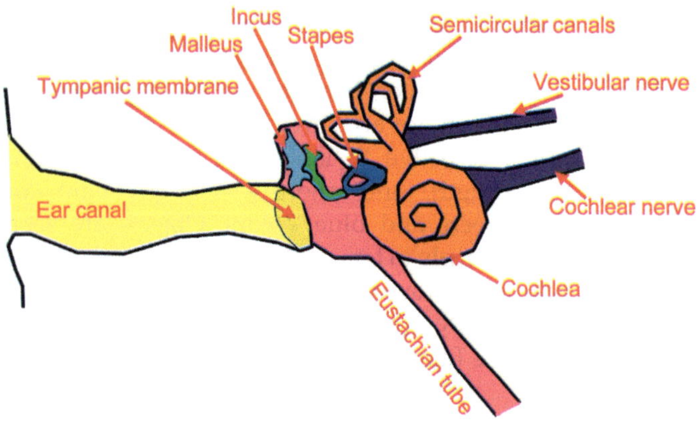

FIGURE 3.5.2 Anatomy of the ear.

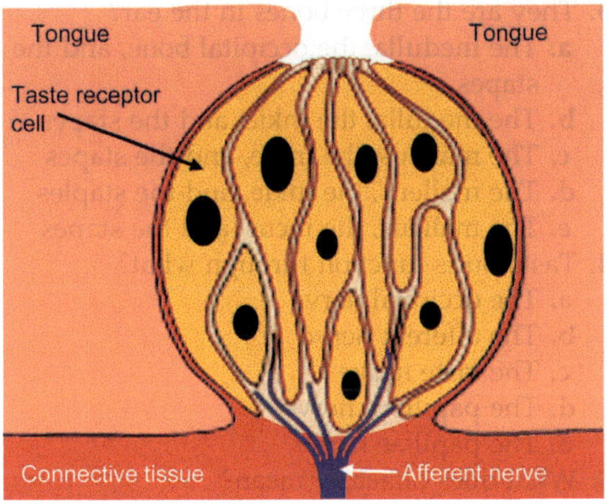

FIGURE 3.5.3 Anatomy of a tongue taste bud or papilla.

SMELLING

Human detection of smell, also called olfaction, is a sense mediated by specialized sensory cells of the nasal cavity. Olfaction occurs when odorant molecules bind to specific sites on the olfactory receptors in the nose. Olfactory cells come together at the glomerulus, a structure that transmits signals to the olfactory bulb, a brain structure directly above the nasal cavity and below the frontal lobe. Olfactory sensory neurons in the olfactory bulb project smells to the brain within the olfactory nerve (cranial nerve I). Olfaction, along with taste, is a form of chemoreception.

TOUCHING

The fifth or final sense is touch (the five senses are sight, sound, taste, smell, and touch). Touch, or somatosensation or mechanoreception, is a sense resulting from neural receptors in the skin including hair follicles. Pressure receptors in the skin respond to variations in pressure. The impairment of the ability to feel is called tactile anesthesia. Paresthesia is a sensation of tingling or numbness that results from nerve damage and may be permanent or temporary.

CHAPTER KEYWORDS

Know these keywords and what each word refers to as well as its function. Read over the list. Are all these keywords familiar?

Afferent nerve
Anterior chamber
Cochlea
Cone
Cornea
Ear canal
Incus
Inner ear
Iris
Lens
Malleus
Mechanoreception
Middle ear
Olfaction
Papilla
Paresthesia
Retina
Rod
Sclera
Semicircular canal
Somatosensation
Stapes
Taste receptor cell
Tympanic membrane
Vestibular nerve
Vitreous fluid

INTERNET REFERENCES

Use these references to find out more.
en.wikipedia.org/wiki/Eye
en.wikipedia.org/wiki/Ear
www.allaboutvision.com/resources/anatomy.htm
en.wikipedia.org/wiki/Taste
en.wikipedia.org/wiki/Olfaction
en.wikipedia.org/wiki/Olfactory_system

CHAPTER QUIZ

Answer these questions. Do you know the answers, holy cow! Check yourself out in this quiz, and prove that you are a humanology wiz.

1. The Iris of the eye serves what function?
   a. It is the diaphragm of eye.
   b. It is the part of the lens of the eye.
   c. It is the protective cover of the eye.
   d. It is the part of the retina that deals with a black and white signal.
   e. It is the part of the retina that deals with a color signal.
2. What is the rod in the eye?
   a. It is the diaphragm of eye.
   b. It is the part of the lens of the eye.
   c. It is the protective cover of the eye.
   d. It is the part of the retina that deals with a black and white signal.
   e. It is the part of the retina that deals with a color signal.
3. They are the three bones in the ear?
   a. The medulla, the occipital bone, and the stapes
   b. The medulla, the inkie, and the stapes
   c. The malleus, the inkie, and the stapes
   d. The malleus, the inkie, and the staples
   e. The malleus, the incus, and the stapes
4. Taste buds function through what?
   a. The occipital nerve
   b. The afferent nerve
   c. The taste nerve
   d. The papillary nerve
   e. The papillae nerve
5. What does olfactory mean?
   a. Smell
   b. Stink
   c. Hearing
   d. Taste
   e. Related to the eye

CHAPTER

3.6

The Liver and Digestive System

DIGESTION

The entire human digestive tract, mouth to anus, is illustrated in Figure 3.6.1. As shown, as food enters the mouth the long digestive process starts. In the mouth, salivary enzyme Amylase cleaves starchy sugars like those in potatoes and rice down to glucose (Table 3.6.1). The enzyme Ptyalin also cleaves these sugars down to the disaccharide maltose. From there food passes down the esophagus to the stomach. In the stomach, which has an acid pH, the enzyme Pepsin cleaves all proteins after the amino acids Phe, Trp, and Tyr. The enzyme Rennin curdles milk by cleaving the milk protein caseinogen into the insoluble protein casein (Table 3.6.1). The acid of the stomach, dilute hydrochloric acid, pH 1.5-3.0, in itself cleaves the amino acids in many proteins and breaks up complex fats and sugars like starch.

After the stomach, food passes into the alkaline duodenum (Figure 3.6.1). Here it meets a group of enzymes produced and secreted by the pancreas as pancreatic juice. This is Trypsin, which cleaves proteins at the amino acid Lys and Arg, and Chymotrypsin, which cleaves proteins at Trp, Tyr, Phe, Leu, and Met (Table 3.6.1). The enzyme Lipase is also produced, which cleaves complex fats and triglycerides into fatty acids. Finally, the enzyme Nucleotidase in the pancreatic juice cleaves RNA and DNA down to their root nucleotides. Food then passes into the small intestine where the enzyme Maltase cleaves maltose generated by the enzyme ptyalin into glucoses. Serial aminopeptidases and carboxypeptidase cleave amino acids in the cut-up proteins one amino acid at a time, starting at the amino terminal and carboxyl terminal of peptides, respectively (Table 3.6.1). At this point, in the small intestine, most proteins have been cleaved to amino acids, most polysaccharides have been cleaved to monosaccharides, most lipid or triglycerides have been cleaved into fatty acids, and most DNA/RNA has been split into root nucleotides. Most of these root molecules are absorbed by the intestines into the blood.

Digestion is, of course, not just enzymatic degradation. It is the mechanical break up of foods, mastication, which starts with mouth action, i.e., chopping up food. Muscular peristalsis waves that continue on in the stomach convert food from a solid mass into syrupy liquid. It is the combination of mechanical muscular digestion and enzymatic digestion that turns solid food into amino acids, monosaccharides, and fatty acids, plus nucleotides, vitamins, and minerals.

Human Physiology, Biochemistry and Basic Medicine
http://dx.doi.org/10.1016/B978-0-12-803699-0.00029-3

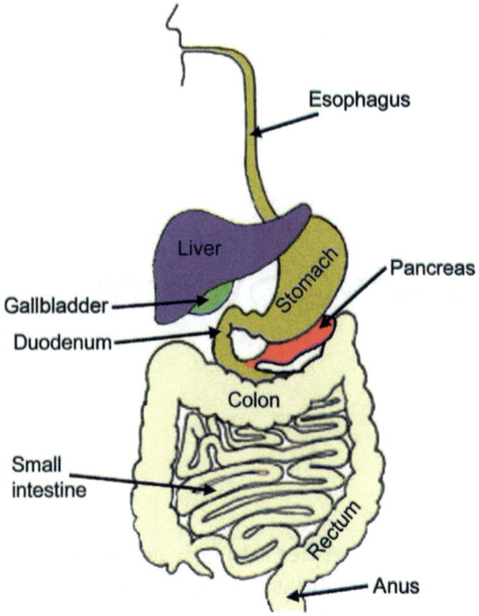

FIGURE 3.6.1 The human digestive system.

TABLE 3.6.1 Hydrolases in the digestive system

Saliva in mouth	
Amylase	Cleaves starchy sugars → glucose
Ptyalin	Cleaves starchy sugars → disaccharide maltose
Stomach	
Pepsin	Cleaves peptides/proteins at Phe, Trp, and Tyr residues
Rennin	Cleaves protein caseinogen → casein
Pancreatic juice in duodenum	
Trypsin	Cleaves peptides/proteins at Lys, Arg residues
Chymotrypsin	Cleaves peptides/proteins at Trp, Tyr, Phe, Leu, and Met residues
Lipase	Triglycerides → fatty acids
Nucleotidase	RNA/DNA → nucleotides
Small intestine	
Maltase	Disaccharide maltose → glucose
Sucrase	Disaccharide sucrose → glucose
Aminopeptidases	Amino terminal exopeptidase (one amino acid at a time)
Carboxypeptidases	Carboxyl terminal exopeptidase (one amino acid at a time)

LIVER

The liver is an essential organ in the digestive system. It has numerous functions, including detoxification of an ingested sustenance, protein synthesis, and production of biochemicals necessary for digestion. Its essential functions include glycogen storage, decomposition of red blood cells, plasma protein synthesis, hormone production, and detoxification.

The functions of the liver are carried out by hepatocytes or liver cells. The liver is thought to be responsible for up to 500 separate functions, usually in combination with other systems and organs. The liver is the site of amino acid synthesis as described in Chapter 1.4. The liver performs multiple roles in carbohydrate metabolism. It is the site of gluconeogenesis, glycogenolysis, glycogenesis, and ketogenesis. The liver is responsible for the mainstay of protein metabolism, synthesis, as well as degradation. The liver is also the site of cholesterol synthesis and lipogenesis or the production of triglycerides and complex lipids or fatty acid derivatives.

In the first trimester of pregnancy in the fetus, the liver is the main site of red blood cell production, a task taken over by bone marrow toward the end of pregnancy. The liver produces and excretes bile, a yellowish liquid, required for emulsifying fats and helping the absorption of vitamin K from the diet. The liver also produces insulin-like growth factor 1 (IGF-1), a protein hormone that plays an important role in childhood growth. The liver is a major site of thrombopoietin production. Thrombopoietin is a glycoprotein hormone that regulates the production of platelets by the bone marrow.

Among the many functions of liver cells is the breakdown of insulin and other circulating hormones. The liver also breaks down or modifies toxic substances in the blood, including most medicinal products or drugs administered in a process called drug metabolism. The liver converts ammonia to excreted urea.

Among the liver's other functions, it stores glucose as glycogen. The liver is responsible for immunological effects, acting as a "sieve" for antigens carried to it via the portal system. The liver produces albumin, a major protein component of blood serum. The liver also functions as a blood reservoir. Large quantities of blood can be stored in reserve in its blood vessels. During cardiac failure, the liver expands, and 0.5-1 liter of extra blood can be supplied by the liver.

CHAPTER KEYWORDS

Know these keywords and what each word refers to as well as its function. Read over the list. Are all these keywords familiar?

Aminopeptidase
Amylase
Carboxypeptidase
Casein
Caseinogen
Chymotrypsin
Colon
Duodenum
Esophagus
Gallbladder
Glycogen
Lipase
Lipogenesis
Maltase
Mastication
Nucleotidase
Pancreas
Pepsin
Peristalsis
Ptyalin
Rectum
Rennin
Small intestine
Triglyceride
Trypsin

INTERNET REFERENCES

Use these references to find out more.
en.wikipedia.org/wiki/Liver
en.wikipedia.org/wiki/Stomach
en.wikipedia.org/wiki/Duodenum
en.wikipedia.org/wiki/Pancreas
en.wikipedia.org/wiki/Digestion
en.wikipedia.org/wiki/Trypsin
en.wikipedia.org/wiki/Pepsin
en.wikipedia.org/wiki/Chymotrypsin
www.innerbody.com/image_digeov/card10-new2.html

CHAPTER QUIZ

Answer these questions. Do you know the answers, holy cow! Check yourself out in this quiz, and prove that you are a humanology wiz.

1. Trypsin cleaves proteins at:
- **a.** Trp, Tyr, and Phe
- **b.** Leu, Met
- **c.** Lys, Arg
- **d.** Tyr, Phe
- **e.** Gly, Ala

2. Proteins are cleaved in the duodenum by:
- **a.** Stomach protease enzymes
- **b.** Duodenum protease enzyme
- **c.** Pancreatic protease enzymes
- **d.** Liver protease enzyme
- **e.** Small intestine protease enzymes

3. Pepsin is an enzyme produced in the:
- **a.** Mouth
- **b.** Duodenum
- **c.** Stomach
- **d.** Pancreas
- **e.** Small intestine

4. The liver stores:
   - a. Glycogen
   - b. Pyruvic acid
   - c. Glycerol
   - d. Glycine
   - e. Sucrose
5. Which of the following functions cannot be attributed to the liver?
   - a. Gluconeogenesis
   - b. Glycogenolysis
   - c. Glycogenesis
   - d. Drug metabolism
   - e. Production of catecholamines

CHAPTER

3.7

The Kidney Filtration System

The kidney filters the blood using millions of parallel glomerulus-tube-collecting duct filtration mechanisms or nephrons (Figure 3.7.1). Each nephron filters blood in parallel to each other. A normal kidney contains 0.80-1.5 million nephrons. The main function of nephrons is to regulate the concentration of water and soluble substances like sodium salts and nitrogen-containing products in blood by filtering the blood, reabsorbing what is needed and excreting the rest as urine. A nephron rids the body of waste, regulates blood volume and blood pressure, controls levels of salts in the blood, and regulates blood pH.

Composed of a glomerulus and the Bowman's capsule, this is the beginning of the nephron. It is the nephron's initial filtering component (Figure 3.7.1). The glomerulus is a blood capillary complex that receives its blood supply from an arteriole of the renal artery. The glomerular blood pressure provides the driving force for water and solutes to be filtered out of the blood and into the nephrons. Approximately 1/5 of all plasma passing through the kidney is filtered into glomeruli. Water passes through the Bowman's capsule into the glomerular filtrate, as do salts, glucose, urea, amino acids, and most smaller proteins (molecular weight <68,000). No erythrocytes or other blood cells enter the glomerular filtrate.

Juxtaglomerular cells, smooth muscle cells in the walls of glomerulus arterioles, secrete into blood an enzyme called Renin. Secretion of this enzyme controls the body's blood pressure. Renin cleaves angiotensinogen a protein secreted by the liver. This produces Angiotensin I. Angiotensin I is cleaved further by lung Angiotensin-converting enzyme to make Angiotensin II, a hormone. Angiotensin II controls contraction of blood vessels increasing blood pressure.

The nephron carries out most of the kidney's functions. These functions concern the reabsorption and secretion of sodium, carbohydrates, and amino acids. Each segment of the nephron has very specialized functions.

In the proximal tubes, the glomerular filtrate is partly reabsorbed. Much filtered water, sodium, glucose, bicarbonate (soluble CO_2), and amino acids are reabsorbed back into the blood stream by the proximal tubes. The loop of Henle is a U-shaped tube that extends from the proximal tubule. It consists of a descending limb and ascending limb. The primary role of the loop of Henle is to concentrate what remains in the glomerular filtrate (Figure 3.7.1).

The distal convoluted tube has a separate function to that of the proximal convoluted tubule. Cells lining the tubule have numerous

Human Physiology, Biochemistry and Basic Medicine
http://dx.doi.org/10.1016/B978-0-12-803699-0.00030-X

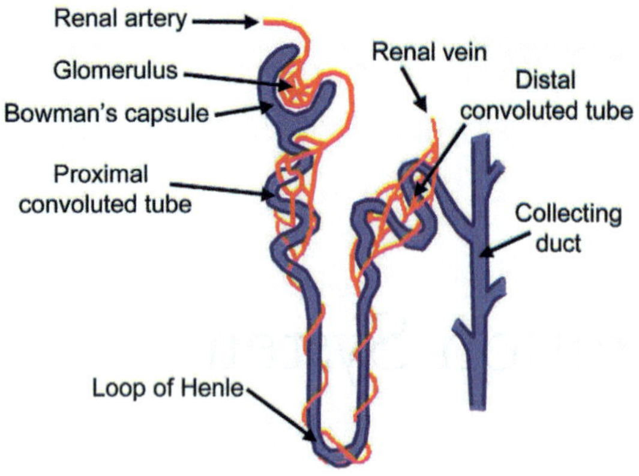

FIGURE 3.7.1 The renal filtration system or nephron.

mitochondria to produce enough energy, or ATP, for active transport to take place. The function of the distal convoluted tubes is general controlled by antidiuretic hormone (ADH), also called vasopressin, an endocrine produced by the posterior lobe of the pituitary gland. Higher blood concentrations of ADH lead to great water reabsorption by the distal convoluted tubes, or leads to blood retention of water. The distal convoluted tubes also control the reabsorption into blood of remaining salts, most notably calcium, phosphate and potassium ions. The distal convoluted tubes can control blood pH by controlling reabsorption of protons (H^+) and ammonia (NH_4^+) ions.

Each distal convoluted tubule delivers its remaining unreabsorbed filtrate to a system of collecting ducts or urine discharge systems. Urine leaves the collecting ducts and empties into the bladder via the ureter (Figure 3.7.1). Excess water, salts, and other excesses, along with urea and a proportion of proteins, enters the bladder. Urine is a liquid by-product of the body. Cellular metabolism makes numerous by-products, many rich in nitrogen, that need clearance from the bloodstream through the urine.

CHAPTER KEYWORDS

Know these keywords and what each word refers to as well as its function. Read over the list. Are all these keywords familiar?

Antidiuretic hormone
Bladder
Bowman's capsule
Collecting duct
Distal convoluted tube
Glomerular filtrate
Glomerulus
Loop of Henle
Proximal convoluted tube
Renal artery
Urea
Vasopressin

INTERNET REFERENCES

Use these references to find out more.
en.wikipedia.org/wiki/Kidney
en.wikipedia.org/wiki/Nephron
www.ivy-rose.co.uk/HumanBody/Urinary/Urinary_System_Nephron_Diagram.php

CHAPTER QUIZ

Answer these questions. Do you know the answers, holy cow! Check yourself out in this quiz, and prove that you are a humanology wiz.

1. Proximal convoluted tubes primarily reabsorb
   - **a.** Water
   - **b.** Potassium
   - **c.** Nucleic acid
   - **d.** Vitamins
   - **e.** Large protein
2. The distal convoluted tubes are
   - **a.** Part of the urinary tract
   - **b.** The site of phosphate and calcium reabsorption
   - **c.** The site of nucleic acid reabsorption
   - **d.** The site of sodium and amino acid reabsorption
   - **e.** Part of the glomerular apparatus

CHAPTER

3.8

Sex and Reproductive Systems: Male

The male reproductive system is in many ways analogous to the female reproductive system. Both the female and male gonads (testis and ovaries) develop from common primordial germ cells and from the parallel Wolffian duct (develops to male reproductive pathway) and Mullerian duct (develops to female reproductive pathway) in the developing fetus.

The male reproductive system is shown in Figure 3.8.1. Spermatozoa are generated in the testis and stored in the epididymis shown in dark red and dark yellow.

The spermatozoa are ejaculated through the vas deferens shown in green. Nutrients are added by the seminal vesicles, Cowper's gland, and prostate gland, and the sperm is discharged through the erect glans penis shown in orange.

HORMONES OF MALE REPRODUCTION

The Leydig and Sertoli cells of the testis are the site of steroid hormone synthesis in the male, testosterone, estradiol, and dihydrotestosterone production (Figure 3.8.6). Steroid production by the testis is controlled by feedback-regulated hormone synthesis in the brain, and by the pituitary gland. The hypothalamus of brain generates pulses of gonadotropin-releasing hormone (GnRH) (Figure 3.8.2). Through a portal circulation, this signals the gonadotrope cells of the pituitary gland, one inch below, to make the glycoprotein hormone luteinizing hormone (LH) and the glycoprotein hormone sulfated hCG (hCG-SO_4). These hormones circulate through the blood and act on Leydig cells in the testis, controlling steroid hormone production (Figure 3.8.2). A low level of LH and miniscule level of hCG-SO_4 are produced. Both glycoprotein hormones have the same biological function as each other, binding the same receptors on testicular Leydig cells. While just a miniscule amount of hCG-SO_4 is produced, it is approximately 100-fold more active than LH, making its production very relevant.

Testosterone and estrogen circulates in the male blood and is picked up by the hypothalamus of the brain. These molecules feed back to the hypothalamus and control the pulse rate of GnRH regulating this complex hormonal control system (Figure 3.8.2).

All three hormones are synthesized from the root steroid cholesterol. LH and hCG-SO_4 control cleavage by cholesterol side chain cleavage

Human Physiology, Biochemistry and Basic Medicine
http://dx.doi.org/10.1016/B978-0-12-803699-0.00031-1

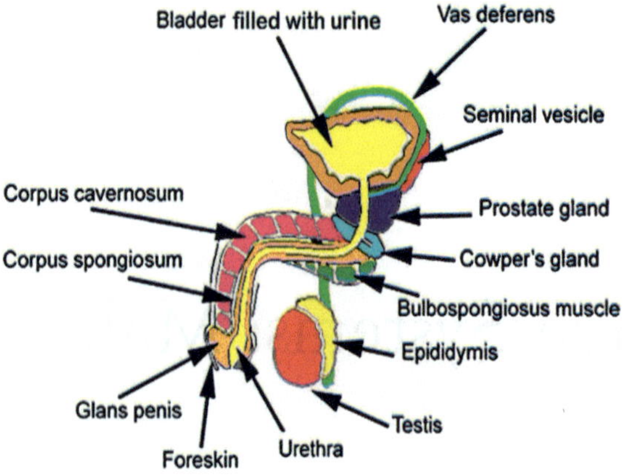

FIGURE 3.8.1 The male reproductive system.

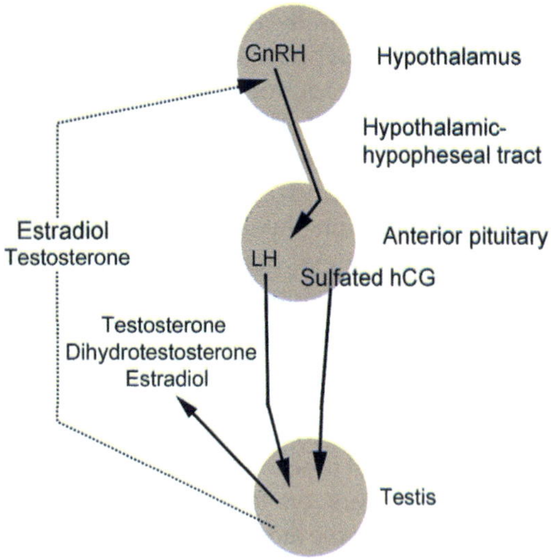

FIGURE 3.8.2 Hormonal control of testicular steroid production.

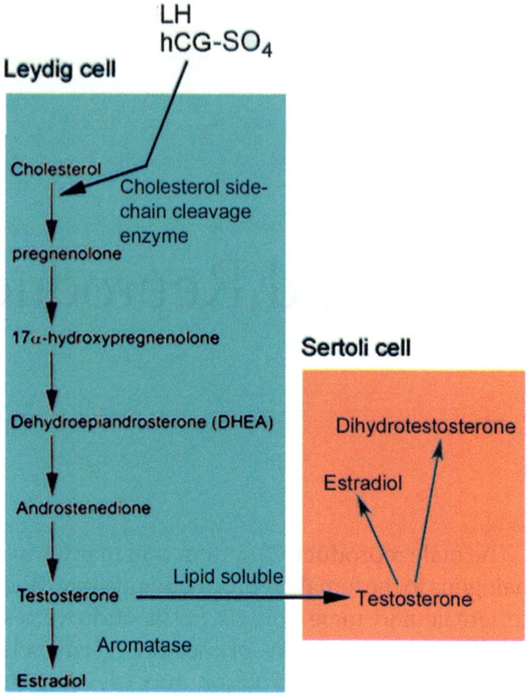

FIGURE 3.8.3 Steroid hormone production in the testis.

enzyme, the first step in all three steroid hormone's synthesis (Figure 3.8.3). Cholesterol is a steroid that is a lipid complex. It is folded to form four carbon loops, three six-membered loops, and one five-membered loop, which form the definition of a steroid. It has 27 carbon molecules having a composition of $C_{27}H_{46}O$. A cholesterol side chain cleavage enzyme cleaves the long fatty acid side chain on cholesterol-making Pregnenolone, a progesterone derivative with only 21 carbons. In steroid synthesis, the carbons are cleaved further to make the androgen dehydroepiandrosterone (DHEA) with only 19 carbons. Finally, testosterone is aromatized and gets an aromatic group on hexagon number one to make estradiol with only 18 carbons (Figure 3.8.3). The 18-carbon estradiol still maintains the 3-carbon hexagonal and 1-carbon pentagonal steroid structure.

All the steroid synthesis up to this point occurs in testicular Leydig cell. Steroid synthesis continues in testicular Sertoli cells (Figure 3.8.3). Dihydrotestosterone is synthesized here. Dihydrotestosterone is a super-active variant of testosterone, with approximately three times the potency.

Testosterone circulates in men, and is a hormone that changes bone density, fat distribution, muscle strength and muscle mass, sex drive, and sperm production in men. If a man loses hormone production of testosterone, multiple dermal preparations are available of replacement hormone that replaces and enriches all these functions. Estradiol in men prevents atherosclerosis, osteoporosis, stroke, and coronary artery disease. It also functions to prevent apoptosis of male sperm cells, maintaining male sperm count. It is understood that estradiol has a profound effect on bone mass in men. It improves arterial blood flow and is involved in human liver and brain function. As such, the estrogen estradiol is critical in men. A small concentration of dihydrotestosterone is produced in men, and has proven functions in the prostate, testes, hair follicles, and adrenal glands.

SPERMATOGENESIS

Spermatogenesis requires the hormones LH and testosterone to be circulating. It also requires the pituitary glycoprotein hormone follicle-stimulating hormone (FSH) to be circulating. FSH is produced by gonadotrope cells in the pituitary aside LH and sulfated hCG-SO_4. All three glycoprotein hormones, LH, FSH, and hCG-SO_4, are produced by gonadotrope cells and are promoted by hypothalamic GnRH. A fourth protein hormone, Inhibin, is produced by testicular Sertoli cell. Inhibin independently feeds back to pituitary gonadotrope cells and controls FSH secretion. This is a small feedback loop with the greater feedback loop, testosterone and estrogen, feeding back to the hypothalamus.

Spermatogenesis is the mechanism by which male germ cells become sperm cells. It is dealt with here in three stages, mitotic proliferation, meiotic division and cytodifferentiation, and packaging of sperm cells. A massive amount of sperm is generated in males; approximately 500 spermatozoa are made per gram of testis per second, or millions of sperms cells each day. Second, seminal fluids are considered, which are ejaculated along with the sperm.

The germ cells of the immature testis are the root cells in the production of sperm, and become activated at puberty. These are the spermatogonial stem cells. Over 36-48 h, these cells undergo multiple divisions or cloning steps, multiplying these cells many times over. If cells undergo 6 divisions, for instance, each cell becomes 64 cells or is multiplied 64-fold (Figure 3.8.4). After 4-5 divisions the final clones become resting primary spermatocytes (Figure 3.8.4).

This mitotic division or cloning occurs in the testicular seminiferous tubules (Figure 3.8.6). During this rapid cloning process, nuclear division is always completed, but cytoplasmic division remains incomplete. As such, all cloned spermatogonia remain linked together through their cytoplasm, joined by a cytoplasmic bridge, which makes spermatogonia strings.

FIGURE 3.8.4 Cloning (mitotic proliferation) of spermatogonia.

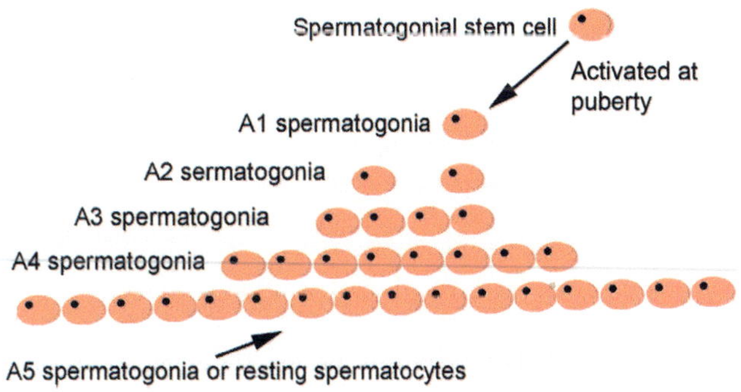

The resting primary spermatocytes now undergo both stages of meiosis (Figure 9.3). All 23 chromosome pairs undergo meiosis. This is the process whereby duplicate pairs of 23 chromosomes, diploid chromosomes, one set coming from the father and one set coming from the mother, are shuffled up and generate one shuffled up half set, or haloid set, of 23 chromosomes. Each diploid cell with 23 pairs of chromosomes becomes during meiosis 4 haploid cells ready for use in sperm and for combining with haploid chromosomes in an egg (Figure 3.8.5). In the process of meiosis, each pair of chromosomes is first duplicated (generating 4 chromatids) in prophase 1-leptotene (Figure 3.8.5), then tied together in prophase 1-zygotene (Figure 3.8.5), and then undergoes interchange of genes between all 4 chromatids or "chiasmata," in prophase 1-diakinesis. Then

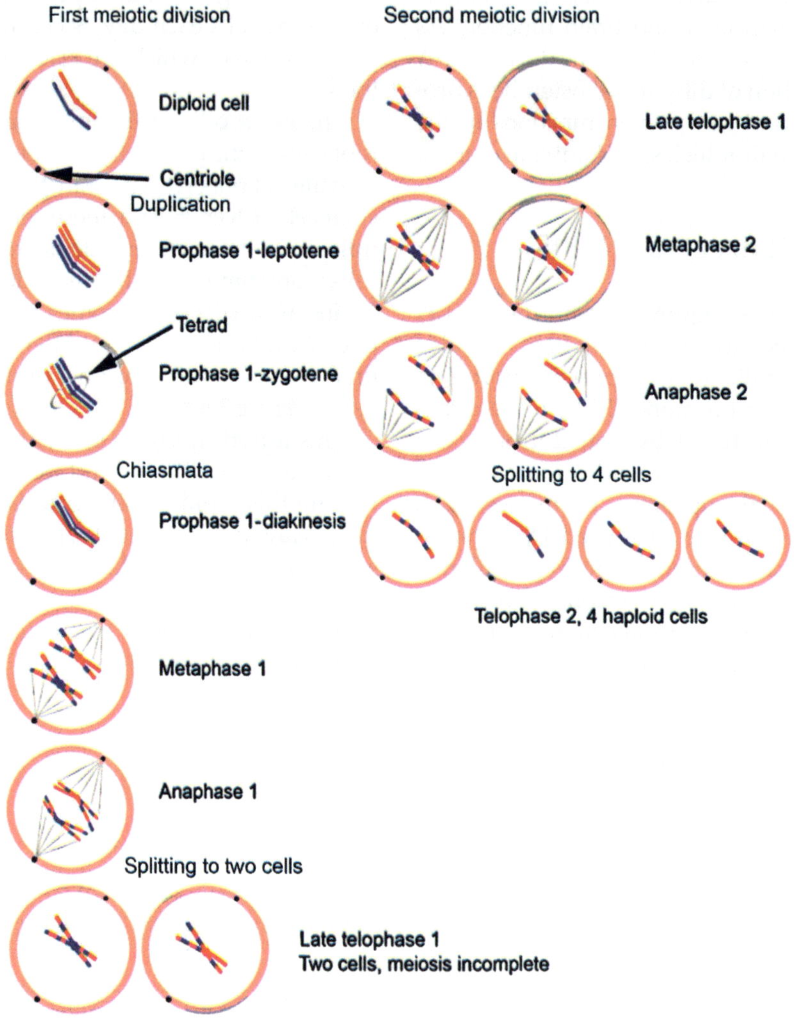

FIGURE 3.8.5 The two stages of meiosis. The example cells have one pair of chromosomes; in reality 23 diploid chromosomes undergo meiosis to make haploid cells.

chromatids are pulled apart in anaphase-1 to make two cells in late telophase 1 (Figure 3.8.5). In Anaphase 2, all 4 chromatids are completely separated and make 4 cells each containing a single or haloid chromatid set telophase 2 (Figure 3.8.5).

The result of meiosis is "early round spermatids." It can be calculated, that if a spermatogonia underwent 6 stages of cloning, it would generate 64 primary spermatocytes, after entering meiosis this would generate 256 early round spermatids. Considering losses due to failures, the actual number may be much less than this.

The most difficult stage of spermatogenesis is the cytoplasmic differentiation and remodeling. This is performed at the seminiferous tubule/Sertoli cell interface (Figure 3.8.6). Round spermatids are reshaped and a tail is added to give the spermatozoon propulsion. In the Golgi apparatus of the reshaped sperm, an acrosome granule forms growing over the nucleus forming a cap-like structure (see Figure 3.8.7). A single fiber flagella or tail is formed that descends from the acrosome and nucleus. The cytoplasm then becomes very squeezed an elongated. A neck or connecting piece is formed by the cytoplasm, and the nucleus and acrosome become flattened (Figure 3.8.7). Nine fibers then form a long tail, forming the spermatozoon. Mitochondria accumulate in the middle piece of the sperm as a power center for tail propulsion. While mitosis or cloning occurs very quickly, these stages of differentiation and packaging of sperm cells are slower, taking 45-50 days.

The flagella of spermatozoon are filled with dense fibers that move with energy, ATP-generated energy in the mitochondria permitting spermatozoa propulsion. A mature spermatozoon is shown in Figure 3.8.7.

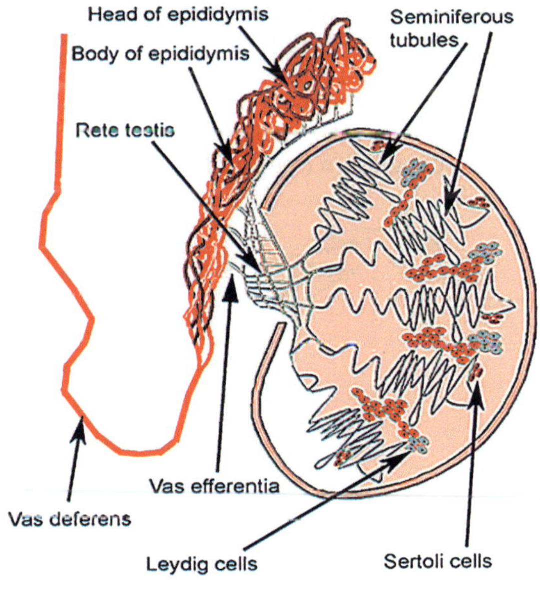

FIGURE 3.8.6 The anatomy of the testis.

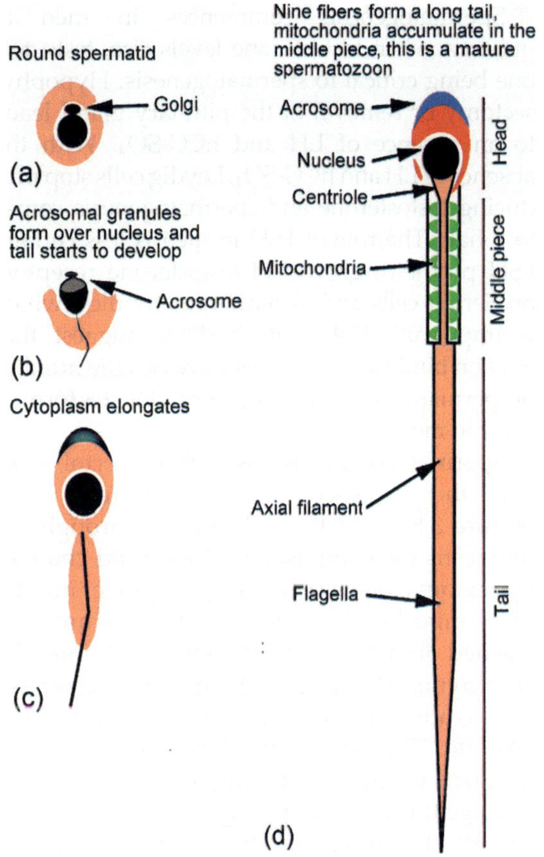

FIGURE 3.8.7 Cytodifferentiation of spermatids to form spermatozoon.

Spermatogenesis commences in men at puberty, when testosterone levels rise, testosterone being critical to spermatogenesis. Hypophysectomy or removal of the pituitary gland leads to an absence of LH and hCG-SO_4. With the absence of LH and hCG-SO_4, Leydig cells stop producing testosterone and spermatogenesis comes to a halt. The role of FSH in men is less certain. FSH promotes growth of testosterone receptors on Sertoli cells and seminiferous tubules, which is important. Data with rodents suggest that FSH on binding Sertoli cells increases the number of spermatogonia or resting spermatocytes formed prior to meiosis.

Spermatozoa are released from Sertoli cells and then washed toward the rete testis (Figure 3.8.6). As the fluid passes through the rete testis the composition of the fluid changes it becomes oxygen rich, needed to power middle piece mitochondria. The spermatozoa are then washed into the vas efferentia and into the epididymis (Figure 3.8.6). In the epididymis, fluid is adsorbed and spermatozoa concentrated 100-fold. The epididymis then adds secretory products to sperm including the sugar fructose. Passage through the epididymis takes 6-12 days and affects spermatozoa behavior. Spermatozoa in the epididymis have acquired what is needed to swim progressively by being supplied with nutrients, fructose, as a sugar fuel. This whole process of activation of spermatozoa is crucially dependent upon stimulation of the epididymis by testosterone.

After storage in the epididymis, spermatozoa enter the vas deferens (Figure 3.8.6). In the absence of ejaculation, spermatozoa seep into the urethra and are washed away.

Ejaculated spermatozoa are carried to the female tract in seminal plasma. The combination of spermatozoa and seminal plasma are called semen. The seminal plasma provides nutritional factors such as the sugars fructose and sorbitol. It also provides an alkaline buffering capacity to neutralize the acidic vagina. The vas deferens first delivers the spermatozoa to the seminal vesicles (Figure 3.8.1). From the seminal vesicles the sperm passes into the prostate gland and then through the Cowper's gland before entering the urethra and being ejaculated (Figure 9.6). In the seminal vesicles, most of the seminal plasma is made. A protein called lipofuscin from dead epithelial cells gives the semen a yellowish color. The seminal vesicles supply the semen with fructose and sorbitol as fuel and an alkaline buffer to cope with the acidity of the vagina. In the prostate gland, semen receives citric acid at a calcium chelator. The semen is supplied with 1% proteolytic enzymes to clear the way of spermatozoa into ovum. The Cowper's gland provides the semen with a mucus lubricant that clears the flow of semen through the urethra.

ERECTION AND EJACULATION

Penis erection can be promoted by psychological stimuli and by erotic imagery. These are developed in the hypothalamus, temporal lobe, and associated areas of the brain. A signal is sent through the spinal cord to the genitalia. Tactile stimulation of the penis can also cause erection. This signals the brain, which in turn signals the genitalia.

The two corpus cavernosa provide most of the erection of the penis (see Figure 3.8.8), aided by the corpus spongiosum (see Figure 3.8.8). Examining

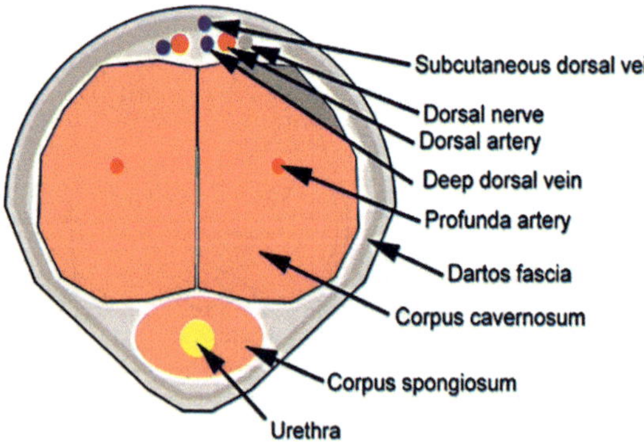

FIGURE 3.8.8 Transverse cut through shaft of penis.

how a flaccid penis becomes fully erect. The nerve signal from the brain is received by parasympathetic and sympathetic nerves on the smooth muscle surrounding the internal pudendal artery and left and right profunda arteries feeding the penis. The smooth muscles contract on the arteries causing dilation and increased blood flow to the two corpus cavernosa and corpus spongiosum of the penis (Figure 3.8.8).

Arteriovenous shunts direct blood to the corpus cavernosa. These events change the corpus cavernosa from low-volume, low-pressure chambers into high-volume, high-pressure chambers, erecting the penis.

As ejaculation approaches the pressures in the corpus cavernosa and corpus spongiosum, or extent of erection, increases further. With further stimulation, a sequence of contraction occurs; the muscle of the prostate, vas deferens, and seminal vesicles expel the spermatozoa and seminal plasma together into the urethra. This is the result of a sympathetic nervous system reaction. Expulsion of the semen, or ejaculation, is achieved by coordinated contraction of the smooth muscles lining the urethra, the bulbocavernosus muscles (Figure 3.8.1), and muscles of the pelvic floor.

With ejaculation the cardiovascular system changes and muscle spasms occur. The orgasm that can occur is associated with sexual tension and arousal. These are suddenly released leading to an intense sensation of pleasure. Males commonly feel tired after intercourse and want to go to sleep. This is because of the amount of exercise involved in intercourse, which leads to sleep.

SEX AND THE HUMAN BRAIN

An area of the brain, the sexually dimorphic nucleus in the preoptic area of the hypothalamus (see Chapter 3.3), is clearly associated with libido and human sexuality. As demonstrated, the sexually dimorphic nucleus in the preoptic area of the hypothalamus is rich in testosterone receptors. In a castrated primate, implantation of testosterone into the sexually dimorphic nucleus in the preoptic area of the hypothalamus restores libido and sexuality. Research suggests that β-endorphins may be released by the sexually dimorphic nucleus in the preoptic area of hypothalamus, and may explain the neural mechanisms underlying sexual behavior. Androgens such as testosterone are the principal stimulators of libido and sexuality in both males and females (testosterone in men 5.5-20 ng/100 ml, in women testosterone 0.20-0.75 ng/100 ml). Androgens act on the brain, on the sexually dimorphic nucleus in the preoptic area of the hypothalamus, promoting libido and sexuality.

The amygdala in the temporal lobe of the brain is approximately the size of an almond. Studies using MRI technology show that the amygdala of the temporal lobe of the brain is involved in human sexual responses. When men are shown pornographic pictures of naked women, or are stimulated by naked women, significant activity can be demonstrated in the amygdala region. It appears that the amygdala is the region of the brain that responds to visual sexual stimulation. This is not surprising since the amygdala is a region that deals with powerful emotions.

Oxytocin, a hormone secreted by the posterior pituitary gland, may also play a key role in male sexuality. Science shows that when a male or female hugs or kisses a loved one oxytocin levels rise. During sexual behavior oxytocin levels also rise. As demonstrated, when oxytocin is injected into the cerebrospinal fluid it causes spontaneous penile erections in males, seemingly reflecting actions on the hypothalamus of the brain. Arletti and Pedersen showed that oxytocin increases sexual receptivity and can counteract impotence. It can somehow make partners cuddle up, and can have a larger role in treatment for infertility. Interestingly, at least two studies have found increases in plasma oxytocin at sexual orgasm in both men and women.

Oxytocin receptors are found in multiple areas of the male brain. These receptors modulate a variety of behaviors, including stress and anxiety, social memory and recognition, sexual and aggressive behaviors, and bonding (affiliation). Research links orgasm, oxytocin, and pleasure sensation in men and women, suggesting that it may be a pleasure hormone in men.

CHAPTER KEYWORDS

Know these keywords and what each word refers to as well as its function. Read over the list. Are all these keywords familiar?

17α-Hydroxypregnenalone
Amygdala
Anaphase
Androstenedione
Aromatase
Bulbospongiosus muscle
Chiasmata
Cholesterol
Cholesterol side chain cleavage enzyme
Corpus cavernosum
Corpus spongiosum
Cowper's gland
Dehydroepiandrosterone
Diakinesis
Dihydrotestosterone
Diploid
Epididymis
Epididymis
Estradiol
Flagella
Follicle-stimulating hormone
Glans penis
Gonadotrope cells
Gonadotropin-releasing hormone
Haploid
hCG-SO_4
Hypothalamus
Leydig cell
Luteinizing hormone
Meiosis
Metaphase
Oxytocin
Pregnenolone
Profunda artery
Prophase
Prostate gland
Rete testis
Seminal vesicle
Seminiferous tubule
Sertoli cell
Sexually dimorphic nucleus
Spermatid
Spermatogenesis
Spermatogonia
Spermatogonial stem cell
Spermatozoa
Sulfate human chorionic gonadotropin
Telophase
Testosterone
Urethra
Vas deferens
Vas efferentia

INTERNET REFERENCES

Use these references to find out more.
www.nlm.nih.gov/medlineplus/ency/article/003707.htm
en.wikipedia.org/wiki/Testosterone
en.wikipedia.org/wiki/Androgen
en.wikipedia.org/wiki/Testicle
www.britannica.com/EBchecked/topic/588769/testis
en.wikipedia.org/wiki/Spermatogenesis
www.uwyo.edu/wjm/repro/spermat.htm
en.wikipedia.org/wiki/Erection
en.wikipedia.org/wiki/Ejaculation

CHAPTER QUIZ

Answer these questions. Do you know the answers, holy cow! Check yourself out in this quiz, and prove that you are a humanology wiz.

1. What are the three steroid hormones made by the testis?
   a. Cholesterol, Testosterone, and Dehydroepiandrosterone
   b. Testosterone, Estradiol, and Dihydrotestosterone
   c. Cholesterol, Estradiol, and Testosterone
   d. Estradiol, Testosterone, and Dehydroepiandrosterone
   e. Pregnenolone, Estradiol, and Testosterone
2. Testosterone is made by:
   a. Sertoli cells
   b. Leydig cells
   c. Gonadotrope cells
   d. Seminiferous tubules
   e. Rete testis
3. Spermatozoa root cells, spermatogonia are:
   a. Round or oval cells
   b. Squarish cell
   c. Cells with a head, middle piece, and tail
   d. Triangular cells
   e. Cell with a flagella or tail
4. Meiosis involves:
   a. Prophase, Anaphase, and Geigaphase
   b. Prophase, Geigaphase, and Terraphase
   c. Metaphase, Picophase, and Femtophase
   d. Picophase, Anaphase, and Terraphase
   e. Prophase, Metaphase, and Telophase
5. What are the two parts of the brain involved in sexual interaction?
   a. Cerebellum and temporal lobe
   b. Hippocampus and thalamus
   c. Temporal lobe and hypothalamus
   d. Cerebellum and basal ganglia
   e. Hypothalamus and Temporal lobe

CHAPTER

3.9

Sex and Reproductive Systems: Female

A woman spends 25-35 years having menstrual cycles. It is during this time that she is fertile and producing active germ cells or ova. The menstrual cycle and germ cell generation is completely regulated by pituitary hormone production. As such, pituitary hormone production is considered first.

ANTERIOR PITUITARY HORMONE PRODUCTION

Three hormones are produced by the anterior pituitary gland gonadotrope cells: the glycoprotein hormone luteinizing hormone (LH), the glycoprotein hormone follicle-stimulating hormone (FSH), and the glycoprotein hormone sulfated human chorionic gonadotropin (hCG-SO_4). All three are released in response to a pulsed signal sent to the pituitary by the hypothalamus of the brain, which releases gonadotropin-releasing hormone (GnRH). The hypothalamus of the brain is just one inch above the pituitary gland. The GnRH signal is sent to the gonadotrope cells through a portal circulation called the hypothalamic-hypophyseal tract (Figure 3.9.1).

The gonadotrope cells of the pituitary release all three glycoprotein hormones in response to pulses of GnRH, LH, FSH, and hCG-SO_4, which all act on ovary cells to promote steroidogenesis. The ovary granulosa and theca cells surround the oocyte or early ova germ cell (Figure 3.9.2). The granulosa and theca cells lie on top of each other surrounding the oocyte (Figure 3.9.2). The granulosa and theca cells each play a separate part in estradiol synthesis. The theca cells are regulated by LH and hCG-SO_4. These hormones control as they do in men a cholesterol side chain cleavage enzyme. Cholesterol, a 27-carbon steroid, is cleaved to pregnenolone and then processed into progesterone, a 21-carbon steroid in theca cells (Figure 3.9.3). Progesterone is further cleaved to make the 19-carbon steroid androgen androstenedione in theca cells. Androstenedione as a lipid or fatty act derivative can diffuse across plasma membranes reaching the adjacent granulosa cells. In the granulosa cells, androstenedione is converted into testosterone, which is cleaved and aromatized to make the 18-carbon aromatized steroid estradiol (Figure 3.9.3).

The menstrual cycle has three phases: the follicular phase or first 12 days of a 28-day cycle, the ovulatory phase or days 12-15 of a 28-day cycle, and a luteal phase or days 16-28 of a 28-day cycle. Primarily estradiol is produced during the follicular phase of the menstrual

http://dx.doi.org/10.1016/B978-0-12-803699-0.00032-3

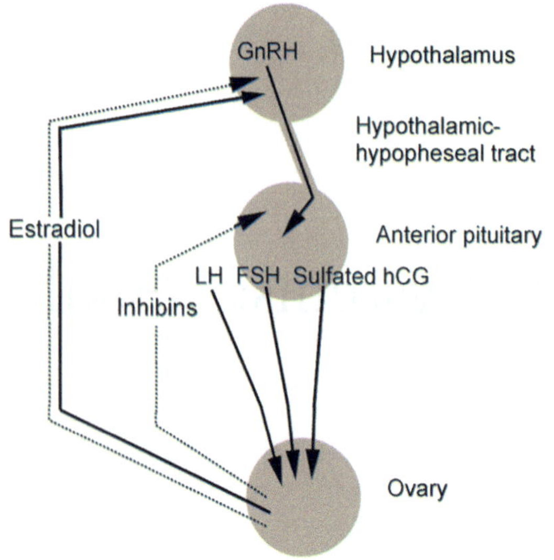

FIGURE 3.9.1 Hormonal control of reproduction in females.

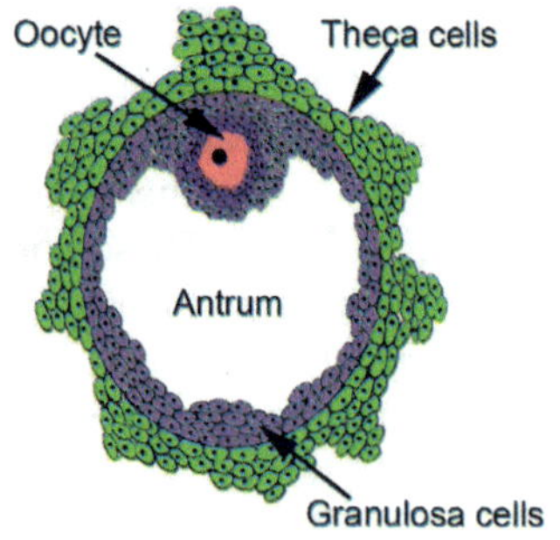

FIGURE 3.9.2 Granulosa and theca cells in the ovary with early oocyte.

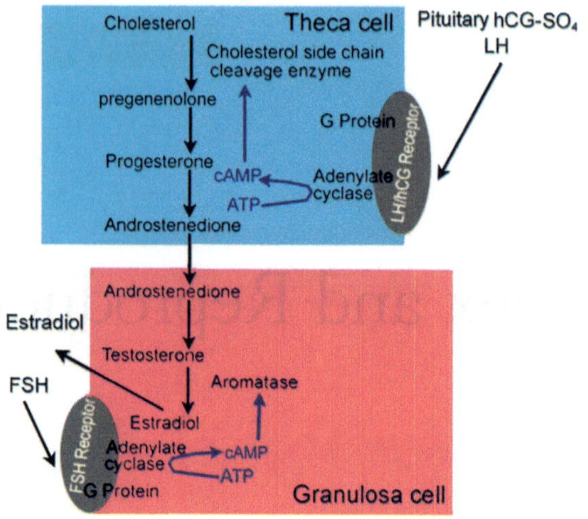

FIGURE 3.9.3 Synthesis of steroid hormones in females.

cycle, estradiol reaching a peak at the start of the ovulatory phase. In the luteal phase, primarily progesterone is produced by corpus luteal cells, the ruin product of granulosa and theca cells, the cells that manage the developing oocyte. Progesterone is made from cholesterol under the control of the hormones LH and hCG-SO_4. It follows the first three steps of the theca cell pathway (Figure 3.9.3).

Women make more steroid hormones than just estradiol in the follicular phase of the menstrual cycle and progesterone in the luteal phase of the menstrual cycle. A small amount of testosterone is made by granulosa cells during estradiol synthesis. This is the hormone that feeds to the sexually dimorphic nucleus in the hypothalamus of the brain and maintains female libido. A total of three variants of estrogens are made. These are estrone, with 18 carbons and one hydroxyl group (abbreviation: E1), estradiol (latin = two), with 18 carbons and two hydroxyl groups (abbreviation E2), and estriol (latin = three) with 18 carbons and three hydroxyl groups (abbreviation E3) (Figure 3.9.4). By far, estradiol (E2) is primarily estrogen detected in the blood. As shown in Figure 3.9.4, estradiol is made by the enzyme aromatase action on testosterone, estrone by aromatase action on androstenedione, and estriol by aromatase action on another androgen, 16α-dehydroepiandrosterone.

During the menstrual cycle, the three estrogens function to promote growth of the endometrial lining of the uterus, in the development of the breast, secondary sexual response, and to promote bone growth and calcium uptake. Progesterone produced during the luteal phase of the menstrual cycle promotes the development

FIGURE 3.9.4 The three estrogens.

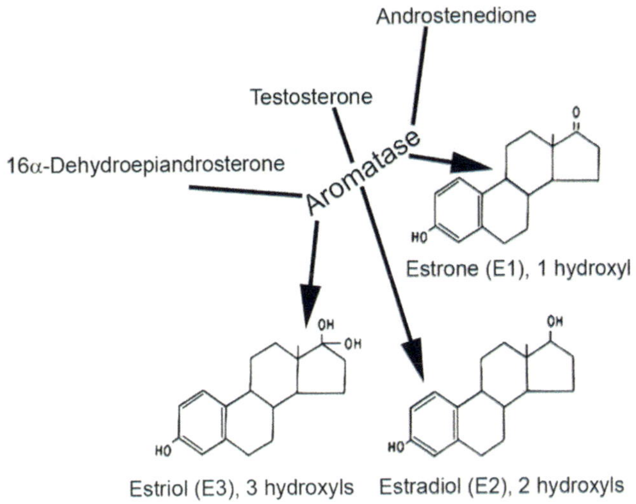

of the endometrium lining, or the glandular receptive tissue for a possible fertilized embryo.

Estradiol has two clear feedback roles (Figure 3.9.1). During the follicular phase of the menstrual cycle, low concentrations of estradiol feed back to the hypothalamus and control GnRH pulses to perfectly balance the GnRH-LH/hCG-SO_4/FSH-Estradiol hormonal system (Figure 3.9.2). Estradiol levels reach a peak or highest blood concentration at the start of the ovulatory phase of the menstrual cycle. At this time, inhibin levels reach a peak. At this time, estradiol levels do not balance; they promote GnRH pulse production, which leads to a peak GnRH signal. Inhibin blocks FSH production, and the result is a peak of LH and hCG-SO_4. These peaks enhance follicular growth and promote ovulation of the ovum.

OOGENESIS

Oogenesis is the pathway whereby women make ova for fertilization by sperm monthly. An average of one active ovum is made by a woman every month. At the beginning of each menstrual cycle, just after the days of menstrual bleeding there are new primitive follicles in the theca-granulosa cell follicular cavity. These are developed by the action of FSH. The competing follicles include surviving atretic follicles from the previous menstrual cycles that never became dominant follicles. There is a total of about 15 follicles in the average woman's ovary. In the first stage of follicular growth, FSH promotes the growth of the primitive follicles and also promotes the recovery of previous menstrual cycle atretic follicles. FSH growth promotion functions through a pathway involving the proteins insulin-like growth factor (IGF)-1 and −2. Follicles start competing in the race to become the largest follicle or the single dominant follicle, the ovum to be ovulated (Figure 3.9.5).

Every follicle is promoted to grow by FSH and IGF-1 and IGF-2, depending upon how large it is or how many FSH receptors it has. In the second stage of growth promotion, every follicle, regardless of its size, produces a set amount of IGF binding protein (IGFBP) (Figure 3.9.5). The set amount of IGFBP-IGF binding protein complex is then destroyed by a set amount of protease or degradative enzyme. In this stage, a large follicle producing IGF over and above that set amount bound by IGFBP fairs the best, with excess IGF-1 and IGF-2

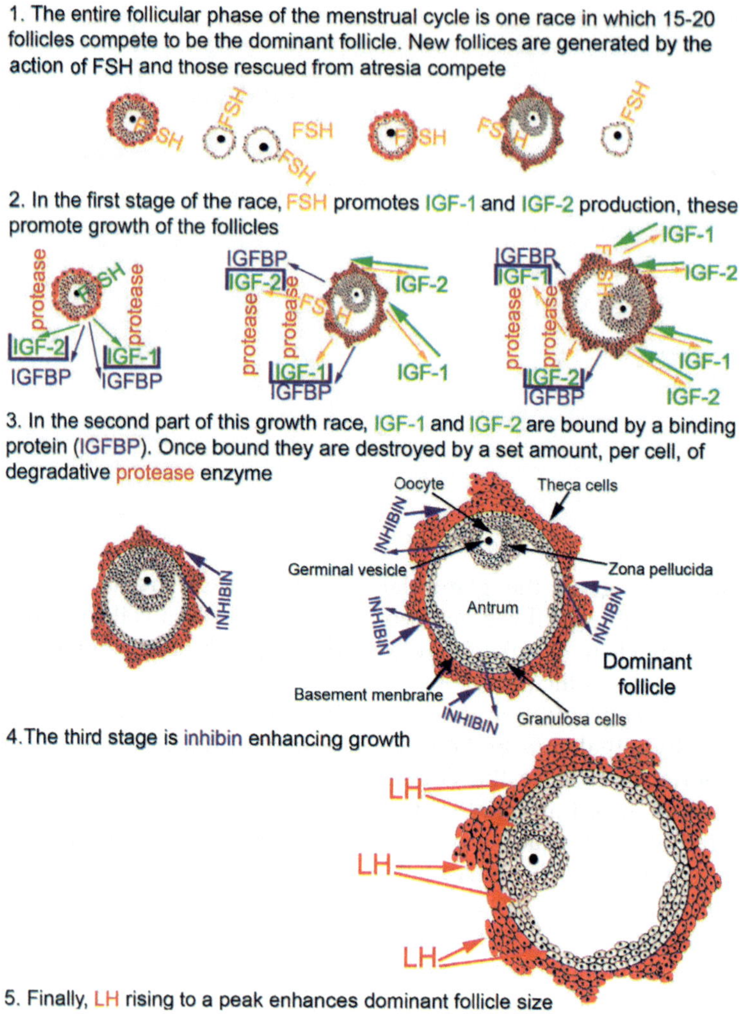

FIGURE 3.9.5 The follicular race. Theca cells are illustrated in light red.

remaining and continuing to promote follicle growth (Figure 3.9.5).

In the third stage of growth promotion, the rising concentration of inhibin toward the end of the follicular phase promotes growth of the follicles. In the fourth stage, in the ovulatory phase of the menstrual cycle, once a dominant follicle or winner of the race becomes obvious, rising LH promotes follicular growth (Figure 3.9.5).

OVULATION

In the follicular phase of the menstrual cycle, estradiol is produced. At the start of the ovulatory phase of the menstrual cycle, estradiol concentration in blood reaches a peak, becoming the highest level in the menstrual cycle. At this same time, production of inhibin, a protein hormone made by granulosa cells, also reaches a high concentration. Inhibin blocks pituitary gonadotrope cells production of FSH. GnRH pulses promotes by the estradiol peak promote all three hormones, LH, hCG-SO_4, and FSH. Inhibin permits separate control of FSH. As such, extreme peak pulses of GnRH in the presence of high concentrations of inhibin will lead to LH and hCG-SO_4 production but not pituitary FSH production, which is blocked by inhibin. This LH and hCG-SO4 production promotes a series of event leading to ovulation of the dominant follicle or ovum.

First, LH and hCG-SO4 promote growth of the dominant follicle (Figure 3.9.5). Second, promotion by LH and hCG-SO4 promotes just the first stage of meiosis in the oocyte (see Figure 3.8.5). The second stage of meiosis is not completed until after fertilization of the ovum (Figure 3.8.5). Third, the thinning of theca and granulosa cell wall close to the oocyte, and formation of a bulge, protrusion, or stigma in the follicular cavity occurs (Figure 3.9.6). This exposes a cell region in the follicular cavity wall for ovulation to occur. Fourth, LH and hCG-SO4 promote production of proteases or degradative enzymes, gelatinase, and collagenase. These enzymes degrade a hole in the thin theca and granulosa cells, the stigma, leading the follicular cavity to pop, or to squirt out the dominant follicle or ovum (Figure 3.9.7).

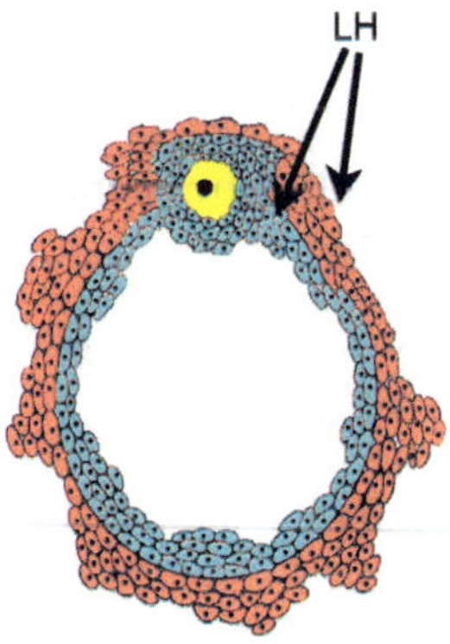

FIGURE 3.9.6 Thinning of the follicular cavity wall and formation of a stigma.

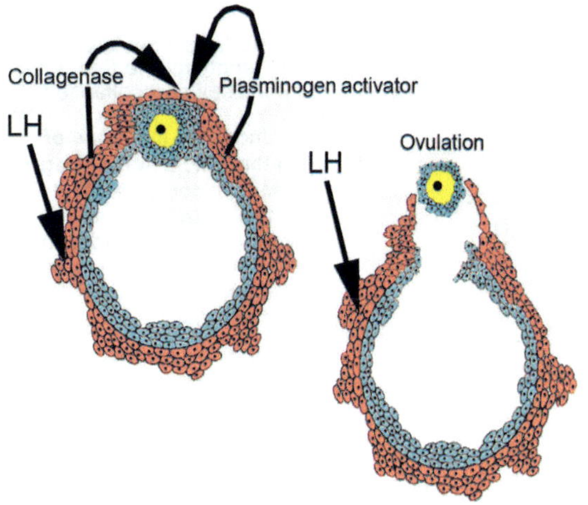

FIGURE 3.9.7 Enzymatic ovulation of the follicular cavity.

The ejected ovum is caught by the fimbriae of the adjacent fallopian tube, and the ovum then starts its journey down the fallopian tube to the uterus.

CORPUS LUTEUM

After ovulation, the theca-granulosa cell cavity is empty. A void is created by ovulation. The empty follicle then collapses into the vacuum space left by the dominant follicle (Figure 3.9.8). Promoted by LH, the follicle hemorrhages and clots among granulosa and theca cells within the follicle. Promoted by LH, luteinization is completed. The granulosa and theca cell hypertrophy and differentiate. The corpus luteum is formed. The corpus luteum takes on a yellow or orangey color (Figure 3.9.8). The granulosa cells form the large lutein cells or the outer crust of the corpus luteum. The theca cells form the small lutein cells or inner mass of cells within the corpus luteum (Figure 3.9.8). The corpus luteum immediately starts producing just progesterone or the product of just the first three enzymatic stages from cholesterol (Figure 3.9.3).

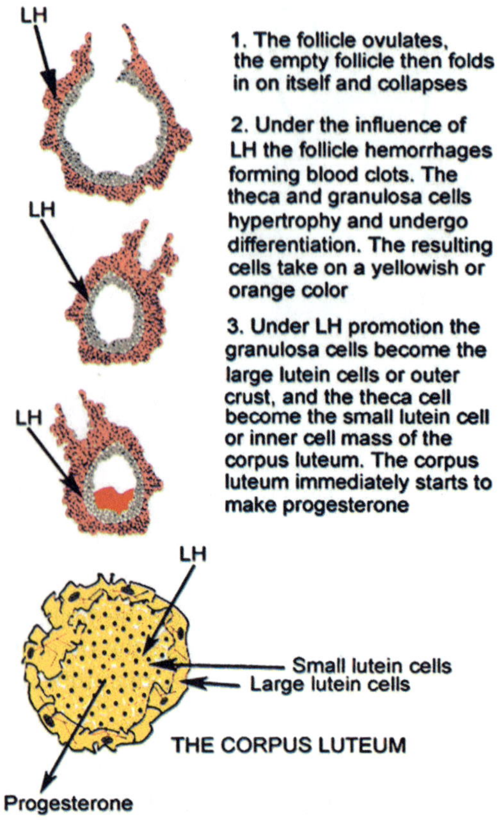

FIGURE 3.9.8 Formation of the corpus luteum after ovulation.

BRAIN AND SEX

As in males, two areas of the brain are clearly involved in female libido and female sexual attractiveness. The sexually dimorphic nucleus in the preoptic area of the hypothalamus is rich in testosterone receptors. In women, the low levels of testosterone produced cause female libido and sexuality (testosterone in men 5.5-20 ng/100 ml, in women testosterone 0.20-0.75 ng/100 ml). Research suggests that β-endorphins may be released by the sexually dimorphic nucleus in preoptic area of hypothalamus, and may explain the neural mechanisms underlying sexual behavior.

The amygdala in the temporal lobe of the brain is approximately the size of an almond. Studies using MRI technology show that the amygdala of the temporal lobe of the brain is involved in human sexual responses. When women are shown pornographic pictures of naked men, or are stimulated by naked men, significant activity can be demonstrated in the amygdala region. This is not surprising since the amygdala is a region that deals with powerful emotions.

CHAPTER KEYWORDS

Know these keywords and what each word refers to as well as its function. Read over the list. Are all these keywords familiar?

Amygdala
Androstenedione
Aromatase
Cholesterol
Cholesterol side chain cleavage enzyme
Collagenase
Corpus luteum
Dominant follicle
E1
E2
E3
Estradiol
Estriol
Estrone
Follicle-stimulating hormone
Follicular phase
Gelatinase
Gonadotropin-releasing hormone
Granulosa cell
$hCG\text{-}SO_4$
Hypothalamus
IGF-1
IGF-2
IGFBP

Inhibin
Insulin-like growth factor binding protein
Insulin-like growth factor-1
Insulin-like growth factor-2
Libido
Luteal phase
Lutein cell
Luteinizing hormone
Oocyte
Oogenesis
Ovulation
Ovulatory phase
Ovum
Pregnenolone
Sexually dimorphic nucleus
Stigma
Sulfated human chorionic gonadotropin
Temporal lobe
Testosterone
Theca cell

INTERNET REFERENCES

Use these references to find out more.
en.wikipedia.org/wiki/Testosterone
en.wikipedia.org/wiki/Androgen
en.wikipedia.org/wiki/Ovary
en.wikipedia.org/wiki/Theca_of_follicle
www.sciencedirect.com/science/article/pii/S1357272505000579
en.wikipedia.org/wiki/Granulosa_cell
en.wikipedia.org/wiki/Estradiol
www.news-medical.net/health/Estradiol-What-is-Estradiol.aspx
en.wikipedia.org/wiki/Estrone
americanpregnancy.org/getting-pregnant/understanding-ovulation/
en.wikipedia.org/wiki/Ovulation
en.wikipedia.org/wiki/Corpus_luteum
www.britannica.com/EBchecked/topic/138543/corpus-luteum

CHAPTER QUIZ

Answer these questions. Do you know the answers, holy cow! Check yourself out in this quiz, and prove that you are a humanology wiz.

1. The three variants of estrogen produced in women are:
   a. Testosterone, estradiol, and estrone
   b. Testosterone, estradiol, and estrone
   c. Estriol, estradiol, and estrone
   d. Estriol, estradiol, and testosterone
   e. Testosterone, estradiol, and androstenedione

2. Testosterone is made in women by what cells?
   a. Leydig cells
   b. Granulosa cells
   c. Theca cells
   d. Sertoli cell
   e. Gonadotrope cells

3. Which of the following is not a promoter of female follicles?
   a. FSH
   b. Inhibin
   c. LH
   d. IGF-2
   e. Follixin

4. In ovulation, meiosis is carried out in the follicular cavity. Which statement is correct?
   a. Just the first meiotic division
   b. Both divisions of meiosis
   c. Just the second division of meiosis
   d. Mitosis is carried out not meiosis
   e. Meiosis only occurs on X and Y chromosomes

5. The corpus luteum only makes
   a. Pregnenolone
   b. Progesterone
   c. Estradiol
   d. Estrone
   e. Androstenedione

CHAPTER

3.10

Fertilization Implantation and Human Pregnancy

OVUM TRANSPORT AND SPERM ACTIVATION

The ovulated ovum is surrounded by a zona pellucida and cumulus cells is caught after ovulation in the fimbriae of the fallopian tube and moved through the fallopian tube to the uterus. The fallopian tube nourishes the moving ovum or moving zygote (fertilized ovum). Secretions from the Peg cells in the fallopian tube provide the ovum nourishment. The smooth muscle wall of the fallopian tubes constricts and contracts, providing propulsion of the ovum through the fallopian tube length. This constriction and contraction is promoted by estrogen, which effectively moves the ovum or zygote from the ovary to the uterus for fertilization (if not already fertilized).

Sperm enters the vagina at coitus (sexual intercourse) and proceeds through the uterus to the fallopian tube. The sperm is propelled by the movement of the flagella or the spermatozoa tail (Figure 3.10.1). If spermatozoa are simply incubated with an ovum in the test tube fertilization will not occur. If spermatozoa are taken from a uterus two hours after intercourse, it somehow becomes capable of fertilization, indicating that something activates sperm in the uterus. As we understand it, when sperm enters an estrogen-primed uterus it is called sperm capacitation and the acrosome reaction occurs. This process super activates the sperm, preparing it for entering the ova vicinity, and activates it for penetration of the ovum zona pellucida and for fertilizing the ovum.

First, a change occurs in sperm capacitation in the movement characteristics of the spermatozoa, to a hyperactive state. The wave-like patterns of the sperm tail or flagella are made stronger, leading to wider amplitude stokes or much more efficient movements, when sperm are swimming in the uterus and fallopian tubes (Figure 3.10.1).

Second, a change occurs in the surface membrane making spermatozoa responsive to signals encountered in the vicinity of the ovum, where an acrosome reaction occurs, leading to increased calcium permeability, leading to a rise in intracellular calcium levels critical to the fertilization process (Figure 3.10.1). This acrosome reaction gives the spermatozoa the changes needed to cross the zona pellucida or shell and

Human Physiology, Biochemistry and Basic Medicine
http://dx.doi.org/10.1016/B978-0-12-803699-0.00033-5

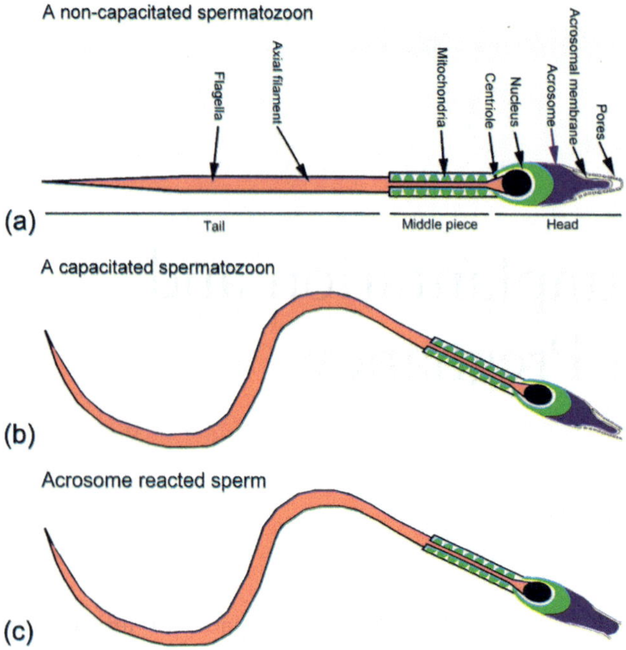

FIGURE 3.10.1 The spermatozoa.

to fertilize the ovum. The acrosome on spermatozoa (Figure 3.10.1) swells and its membrane fuses with the membrane of the spermatozoa. An agent secreted by the zona pellucida of the ovum, called ZP3, seemingly activates this acrosome reaction.

FERTILIZATION

Fertilization is a complex process involving a single sperm penetrating an ovum and the two haploid sets of chromosomes combining to form a fertilized diploid gamete or zygote (Figure 3.10.2). Many things can go medically very wrong. In the case of a complete hydatidiform mole, for instance, the ovum lacks a haploid set, and the haploid product of a single sperm can be duplicated, or two sperm may fertilize an ovum generating a diploid gamete. This oddity, however, generates being a comprising placenta only. In another incidence called a partial hydatidiform mole, the ovum may fail to block multiple sperm fertilization, and a triploid being may be generated, comprising a chopped mixture of placenta and fetal parts.

In 18 of the 50 U.S. states pregnancy or new life is defined as starting with fertilization. Considering the fertilized gamete is in many respects a dying being, with no food source or oxygen supply, considering fertilization complications and that approximately half of all fertilized embryos may fail to implant properly into the uterus and may perish, it is hard to consider that human life starts at fertilization. Life or pregnancy seemingly starts after successful implantation of pregnancy, or after a being has an oxygen, food and blood source, and becomes a flourishing, growing being.

After the acrosome reaction in spermatozoa the new exteriorized membranes are covered in proteolytic or degradative enzymes, equipping then to penetrate the ova's shell or zona pellucida. These then digest or erode a path through the ovum's zona pellucida through which the spermatozoa passes into the ovum. Penetration through the zona pellucida takes approximately

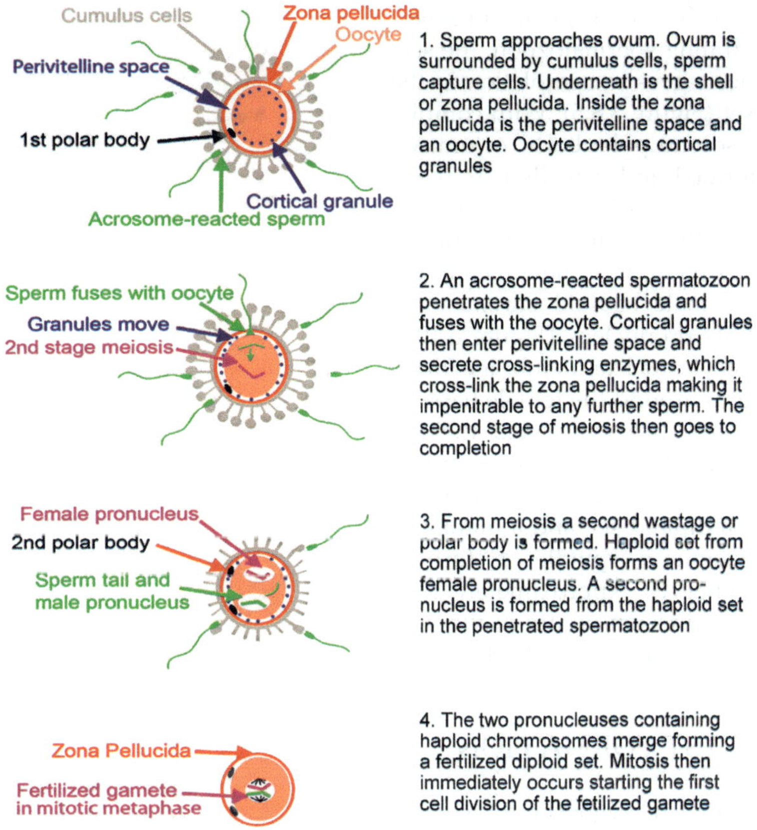

FIGURE 3.10.2 The ovum (oocyte) fertilization process.

5-20 min. The penetrated head of the spermatozoa then fuses with the plasma membrane of the oocyte (Figure 3.10.2, panels 1 and 2).

Only spermatozoa that have gone through the acrosome reaction are capable of permeating the zona pellucida and are capable of fusing with the oocyte plasma membrane. Calcium from the spermatozoa is required for the fusing process. Once fusing occurs, the spermatozoa stops all movement.

As soon as fusion occurs with the oocyte plasma membrane the oocyte becomes a fertilized gamete or zygote. The zygote must quickly prevent other spermatozoa from also trying to fertilize it. Then, it must complete the second stage of meiosis in order to have an active female haploid set of chromosomes to combine with. These critical stages take approximately 20 hours to complete. In the follicle in the ovary only the first stage of meiosis occurred, terminating in the second meiotic metaphase-2 (Figure 3.8.5). The elevated calcium in the spermatozoa generated in the acrosome reaction cause the cortical granules, in the perivitelline space of the ovum after a spermatozoon penetrates the zona pellucida (Figure 3.10.2, panel 2), to fuse together. Upon fusing, the cortical granules release their contents into the ovum perivitelline space. Among their contents are enzymes, which link the glycoproteins within the zona pellucida shell together, cross-linking them, making the zona pellucida impenetrable to further spermatozoa.

In response to the intracellular calcium generated by the penetrating spermatozoon, the second stage of meiosis goes to completion in the female haploid set (Figure 3.8.5). With completion of meiosis a second polar body, or unused haploid set, is formed and extruded from the oocyte (Figure 3.10.2, panel 3). After the spermatozoa have entered the oocyte and the nuclear membrane has broken down, a new membrane forms around the male haploid chromosome set, forming a male pronucleus (Figure 3.10.2, panel 3). A new oocyte nuclear membrane forms a female pronucleus around the female haploid chromosome set that has completed meiosis (Figure 3.10.2, panel 3). The pronuclear membranes then break down, the parental haploid sets mix to make a diploid set, and a mitotic or cell division metaphase spindle is formed (Figure 3.10.2, panel 4). Approximately 24 h later mitosis or cell division occurs in the new zygote leading two zygote identical cells.

BLASTOCYST FORMATION

The newly fertilized zygote now spends approximately six or seven days growing and differentiating while traveling from the fertilization site in the uterine cavity or fallopian tube to the uterus for implantation. During this time, it may develop from a one-two cell zygote into a 32-64 cell blastocyst. Growth, however, may be very much restricted by the firmness of the zona pellucida shell that was originally surrounding the ovum (Figure 3.10.3). Indeed, cells differentiate, but because of the zona pellucida, which is shed prior to implantation, cells get smaller and smaller (Figure 3.10.3). At the Morula stage, 3 days and 16 cells, a first differentiation of identical cells occurs, separating cells destined to be placenta (chorion cells) and cell destined to be the fetus (amnion cells) (Figure 3.10.3). After this, multiple further differentiations occur in placenta and fetal tissues to form the implantable blastocyst (Figure 3.10.3).

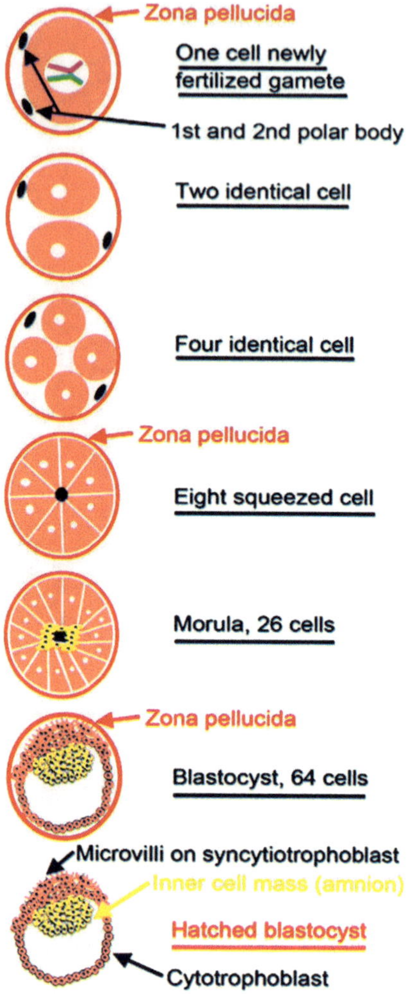

FIGURE 3.10.3 Development of the blastocyst, amnion cells show in yellow.

The zona pellucida also functions in pregnancies to prevent the conceptus from coming apart. If the conceptus does break apart forming two groups of cells or multiple groups of cells, monozygotic (coming from one zygote) twins or multiple pregnancies result. These are genetically identical twins. Generally, monozygote twins come from cells that have split prior to 12 days after fertilizations. After 12 days, cells are well differentiated, and monozygotic twinning due to cells breaking off leads to conjoint or Siamese pregnancies sharing organs or parts. Non-identical twinning only occurs when multiple dominant follicles or ova are ovulated and fertilized by multiple spermatozoa.

On days 6 or 7 after fertilization the embryonic blastocyst hatches or loses the zona pellucida. This exposes some placental cells (shown in Figure 3.10.3 in light red). Cells exposed include placental cytotrophoblast cells and syncytiotrophoblast cells (fused cytotrophoblast cells) with microvilli or protrusions. These placental cells initiate pregnancy implantation.

IMPLANTATION

Implantation is the process whereby the blastocyst buries itself deep inside the uterus, approximately one-third of the way through the thickness of the uterus. The blastocyst first has to be caught by the endometrial lining of the uterus, then pass the endometrial lining and bury itself inside the myometrium or uterine muscle wall.

Before discussing implantation and pregnancy it in necessary to first introduce the hormone human chorionic gonadotropin (hCG), and the associated but very different molecules, the autocrine (acts on cells that produce the molecule) hyperglycosylated hCG (h-hCG). Both molecules are glycoprotein hormones such as pituitary gonadotrope cell luteinizing hormone (LH), follicle stimulating hormone (FSH), and sulfated human chorionic gonadotropin ($hCG\text{-}SO_4$). As a group these molecules are called glycopeptins. These are large molecules comprising a shared common 92 amino acid alpha-subunit (coded on chromosome 6) and a variable length beta-subunit (8 back-to-back copies of hCGß gene on chromosome 19). Each contains strings of sugar side chains making them a glycoprotein that comprise up to 39% of the molecular weight of the molecule. Blastocyst cells and placental cells may be singular as cytotrophoblast cells, or fused cells, 3 to 50 cells fused together as syncytiotrophoblast cell. Syncytiotrophoblast cells make the molecule hCG, and place short sugar side chains on this structure. This is the hormone hCG. Unfused cytotrophoblast cells make the amino acid structure hCG and put large sugar side chains on it (39% of molecular weight). This is the autocrine hyperglycosylated hCG. As found the hormone hCG and the autocrine hyperglycosylated hCG have very different functions.

Evidence shows that as the blastocyst approaches the endometrial lining the zona pellucida is lost or hatched, and the syncytiotrophoblast on the blastocyst are exposed. These produce the hormone hCG, which signals to the endometrium across the uterine cavity space that a blastocyst is coming. These syncytiotrophoblast cells with microvilli (Figures 3.10.3 and 3.10.4) then attach to a glycocalyx, protrusions on the uterine endometrial lining (Figure 3.10.4 and 3.10.5), attaching the blastocyst embryo to the uterus.

Cytotrophoblast cells exposed on the blastocyst produce the autocrine h-hCG. This promotes invasion of the uterine epithelium lining and underlying stromal tissue, in finger-like manners (Figure 3.10.5), or the digging of hole in the uterine surface in an invasive cancer-like manner. This invasion effectively implants the entire blastocyst in the uterus (Figure 3.10.5).

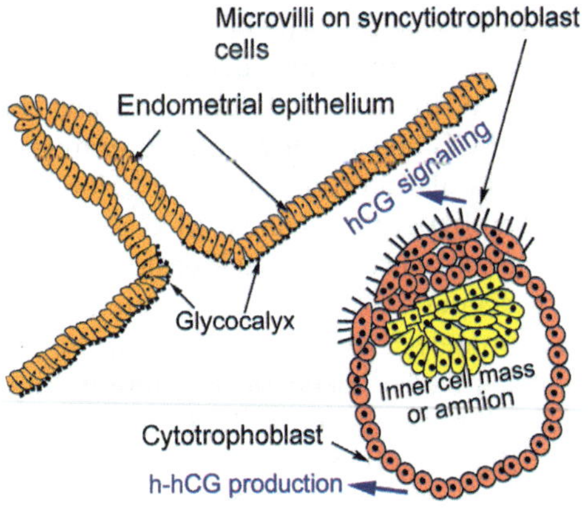

FIGURE 3.10.4 Attachment to the endometrial lining.

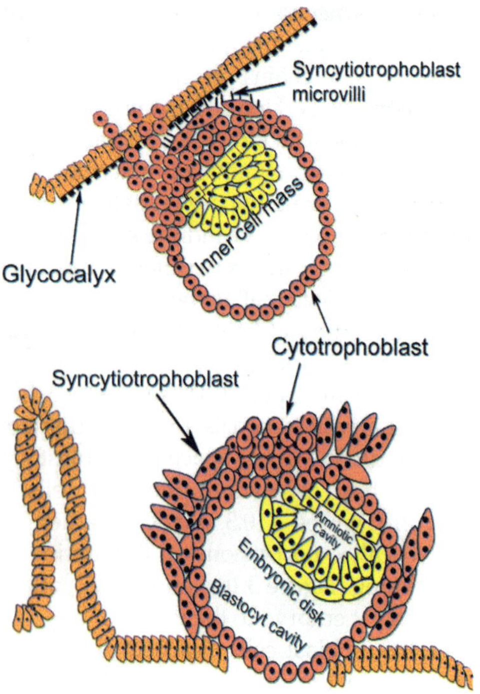

FIGURE 3.10.5 Blastocyst implantation, initial finger-like invasion (upper panel).

Once implanted, the blastocyst differentiates its cytotrophoblast cells, by fusion to syncytiotrophoblast cells (Figure 3.10.5). The blastocyst at 4.5 weeks of gestation is a mixture of invasive cytotrophoblast cells and hCG-producing syncytiotrophoblast cell. The hCG produced enters the maternal circulation, driving progesterone production by the maternal ovarian corpus luteum (hCG takes over corpus luteum progesterone stimulation from pituitary LH and hCG-SO_4). The hyperglycosylated hCG made by the cytotrophoblast cells drives invasion of the blastocyst deeper into the uterus. Eventually, by approximately 5 weeks of gestation, the blastocyst has invaded itself deep enough into the decidua for the uterine epithelium to seal itself closed (Figure 3.10.5).

Research now shows that pregnancy failures, biochemical pregnancies or pregnancies that fail after 2-4 weeks, miscarriages of pregnancies, and ectopic pregnancy are all do improper blastocyst implantation of pregnancy. Apparently, these failed pregnancies can be attributed to insufficient h-hCG production.

PREGNANCY

Once the blastocyst has implanted the cytotrophoblast cells mostly become converted to syncytiotrophoblast cells and hCG the pregnancy hormone is produced at high concentrations and gets to work. hCG promotes maternal corpus luteal progesterone production until approximately 8 weeks of gestation when the syncytiotrophoblast cells start to produce progesterone themselves. As a hormone, hCG has multiple other important functions during the duration of pregnancy. As a circulating hormone it suppresses uterine contractions to take the pregnancy to term. hCG also suppresses maternal macrophage attack of placental and fetal tissues as being maternally-foreign tissues. hCG also promotes the continuous fusion of newly generated cytotrophoblast cell to syncytiotrophoblast cells with 3-50 nuclei. Most of all, the hormone hCG works in unison with the autocrine h-hCG to establish hemochorial placentation, the process whereby the mother passes oxygen, glucose, and nutrients to the developing fetus.

H-hCG functions during pregnancy to promote continuous placental growth, to accommodate the developing fetus. This function is needed for fetal growth for hemochorial placentation or the fetal nutrition system. Hemochorial placentation is a complex but very efficient mechanism of fetal nutrition. It takes from implantation until 10 weeks of pregnancy for the placenta to build this efficient system, which function from 10-40 weeks of pregnancy.

Early in pregnancy the placenta forms villous placenta tissue. This is comprised of star-shaped containers formed from cytotrophoblast cells that fuse to form a skin or fused syncytiotrophoblast

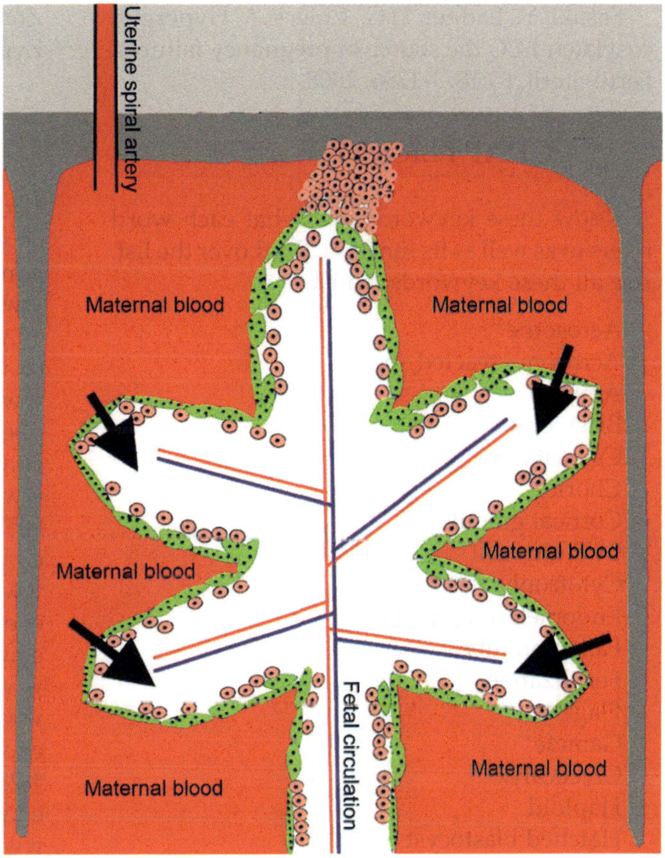

FIGURE 3.10.6 Hemochorial placentation. Shown in light red is cytotrophoblast and in green are syncytiotrophoblast, villous placenta tissues.

cells (Figure 3.10.6). Growth of the cytotropho blast tissue is promoted by the autocrine h-hCG, and fusion to syncytiotrophoblast is promoted by hCG. hCG also promotes the uterine arteries to grow to meet these placenta tissue containers. Inside these containers, under the promotion of the hormone hCG an umbilical circulation linking the fetus to the villous placenta tissue develops. The villous star-shaped containers form inside larger tanks that connect to the uterine arteries. This forms hemochorial placentation (Figure 3.10.6). The average placenta of pregnancy develops 4-6 of these tank systems. These larger tanks fill with maternal blood. Under pressure, the nutrients pass across syncytiotrophoblast cells into the inside of the smaller container. Here the nutrients pass into the umbilical circulation (Figure 3.10.6). This system very efficiently passes oxygen, glucose, and other nutrients from the mother to the fetus.

During pregnancy the fetus develops and is delivered at approximately 40 weeks of gestation.

Hyperglycosylated hCG was a molecule discovered by Laurence Cole PhD, an author of this book.

Cole, L.A. The O-linked oligosaccharides are strikingly different on pregnancy and choriocarcinoma hCG. J. Clin. Endocrinol. Metab., 65:811-813, 1987.

Laurence, A. Cole PhD's laboratory also discovered that hyperglycosylated hCG control pregnancy implantation and were the source of pregnancy failures.

Sasaki, Y., Ladner, D.G., Cole, L.A. Hyperglycosylated hCG the source of pregnancy failures, Fertil Steril, 89:1871-1786, 2008.

CHAPTER KEYWORDS

Know these keywords and what each word refers to as well as its function. Read over the list. Are all these keywords familiar?

Acrosome
Acrosome-reacted sperm
Amnion
Blastocyst
Capacitation
Chorion
Cortical granule
Cumulus cell
Cytotrophoblast
Endometrial epithelium
Fallopian tube
Fertilization
Flagella
Gamete
Glycocalyx
Haploid
Hatched blastocyst
hCG
Hemochorial placentation
H-hCG
Hydatidiform mole
Hyperglycosylated hCG
Implantation
Metaphase
Microvilli
Mitosis
Morula
Oocyte
Ovum
Peg cells
Perivitelline space
Polar body
Pronuclear membrane
Pronucleus
Spermatozoa
Syncytiotrophoblast
Zona pellucida
Zygote

INTERNET REFERENCES

Use these references to find out more.
en.wikipedia.org/wiki/Human_chorionic_gonadotropin
www.hcglab.com/
en.wikipedia.org/wiki/Capacitation
en.wikipedia.org/wiki/Acrosome_reaction
www.embryology.ch/anglais/dbefruchtung/akrosom02.html
en.wikipedia.org/wiki/Fertilisation
www.britannica.com/EBchecked/topic/205305/fertilization
en.wikipedia.org/wiki/Morula
www.embryology.ch/anglais/evorimplantation/furchung01.html
en.wikipedia.org/wiki/Blastocyst
www.oneatatime.org.uk/330.htm
www.ncbi.nlm.nih.gov/pubmed/22386126
en.wikipedia.org/wiki/Implantation_%28human_embryo%29
en.wikipedia.org/wiki/Placentation
www.ncbi.nlm.nih.gov/pubmed/19560
www.placentajournal.org/article/S0143-4004%2810%2900222-5/abstract212

CHAPTER QUIZ

Answer these questions. Do you know the answers, holy cow! Check yourself out in this quiz, and prove that you are a humanology wiz.

1. Before sperm fertilizes the ovum it must go through:
   a. Just the acrosome reaction
   b. The acrosome reaction and capacitation
   c. The acrosome reaction, capacitation, and uterospermiogenesis
   d. Just capacitation
   e. The acrosome reaction and uterospermiogenesis

2. Just one sperm and not two sperm may penetrate the ovum, which is due to:
   a. Space for the zonal pellucida to fit two sperm
   b. Cortical granules produce enzymes that seal the zona pellucida
   c. The perivitelline space limitation
   d. The first sperm seals the zona pellucida
   e. The zona pellucida can only be penetrated by one sperm
3. When does a zygote first split into amnion and chorion cells?
   a. At the first cell division
   b. At the blastocyst stage
   c. At the morula stage
   d. At the 8-cell stage
   e. At the 32-cell stage
4. Implantation occurs:
   a. 10 days of after fertilization
   b. 2-3 days after fertilization
   c. 6-7 days after fertilization
   d. On the day of fertilization
   e. 14 days after fertilization
5. Hemochorial placentation normally starts at around:
   a. 4 weeks of pregnancy
   b. 5 weeks of pregnancy
   c. 7 weeks of pregnancy
   d. 9 weeks of pregnancy
   e. 10 weeks of pregnancy

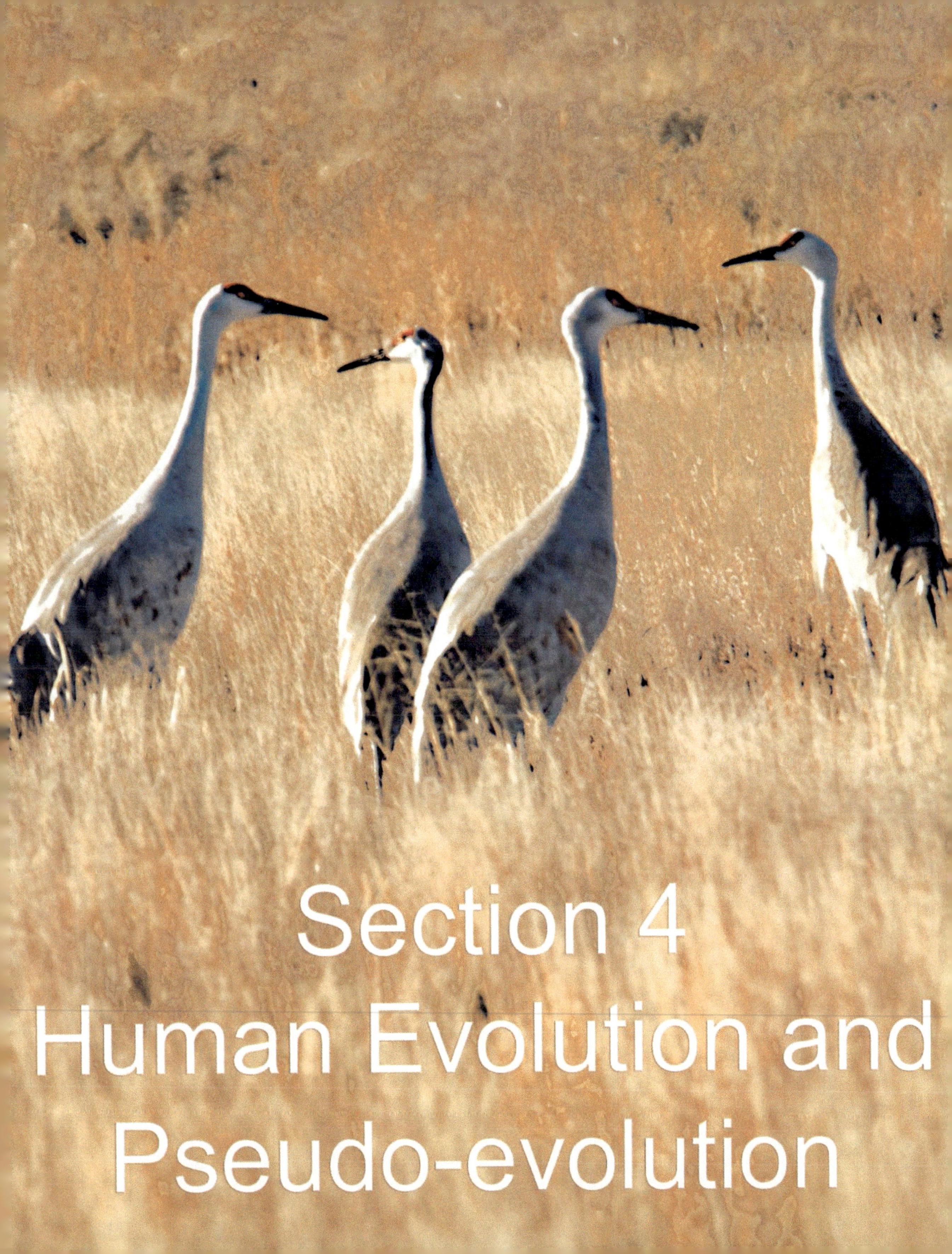
Section 4
Human Evolution and
Pseudo-evolution

CHAPTER

4.1

Human Evolution from Primates

GENERAL EVOLUTION

Charles Darwin sailed to the Galapagos Islands in the South Pacific Ocean on HMS Eagle in 1831. He was fascinated by the islands. Particularly noting that the giant tortoises had different shells on each island, permitting their necks to rise to reach higher plants, or to extend forward to take different foods. Differences were also noted among finches and how their beaks varied on different islands to eat and peck at different foods. In 1859, he wrote the Origin of the Species. He proposed that species adapted by natural selection to their environment, that species varied at childbirth, and that if an adaptation occurred making the species stronger, or providing better food sources, that the adaption was adopted. If an adaption did not benefit a species, it came to an end. This was termed natural selection.

To illustrate the process of natural selection, consider the example of finches on the Galapagos Islands. A mother may have approximately 20 baby finches each year. Normally, just a small number of these 20 babies survive childhood. One baby has a protein mutation that leads to it having a harder beak. This finch is able to break nuts better with its harder beak. As such, it can find and consume more nuts, its principal food. This baby survives and had offspring. A consequence of this mutation that may evolve in future generations is the hard beak and better availability of food. After 20 years, generations of finches, thousands of birds, may have a harder beak.

There are at least five kinds of DNA mutation that can take place, each causing sequence changes or pushing codons to frame shift or describe a nonsense sequence. There are point mutations, caused by chemicals or malfunction of DNA replication, in which an exchange occurs of a single nucleotide for another. There are insertion mutations in which one or more extra nucleotides are inserted inside or added to the DNA sequence. There are deletion mutations in which one or more nucleotides are removed from the DNA sequence, and amplification mutations in which multiple copies of entire regions of DNA are made. There are also Chromosomal Translocations in which interchange of DNA sequences occurs between different chromosomes.

There are four basic causes for mutations: There are spontaneous mutations due to molecular decay; mutations due to an error-prone replication and mutations due to error in DNA

Human Physiology, Biochemistry and Basic Medicine
http://dx.doi.org/10.1016/B978-0-12-803699-0.00004-9

translation; mutations can be caused by errors during DNA repair; and mutations can also be caused by mutagens, radiation and ultraviolet, chemical agents, or contaminants in foods.

Putting all the facts together, a finch, or any other species, may have DNA errors in the reproductive processes. This may cause some simple point mutations. As in the example described above, the DNA codon AAA may mutate to AAC. It codes for the amino acid Asparagine instead of Lysine coded for by AAA. This leads to a structural change in the coded protein. The protein change may be critical for the finch's beak, for example. The finch's beak may, for instance, become hard because of the protein structure change, if this gives the finch an advantage, such that it can eat better or possibly digest nuts better. Then the finch gains strength and gains advantage. By this pathway, Darwin's natural selection, new species evolve and species adapt. As illustrated in Figure 4.1.1, evolution explains the appearance of all species.

Through the mutation mechanism and natural selection, single-cell species became multiple-cell species (Figure 4.1.1). Fish evolved into amphibians, amphibians into reptiles, and reptiles evolved into mammals. Mammals have

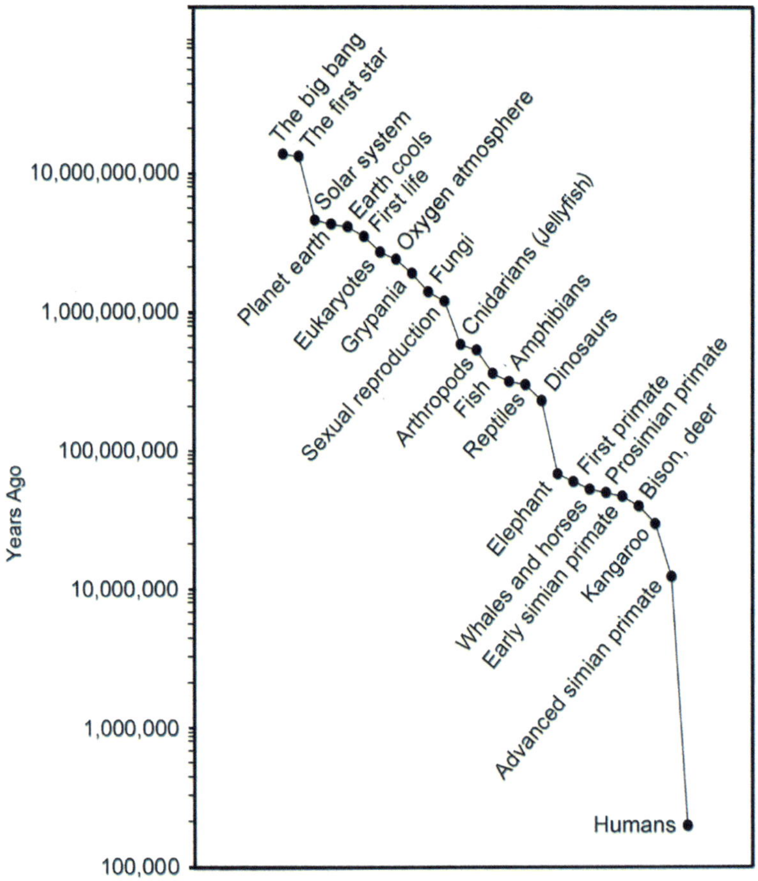

FIGURE 4.1.1 Early physics and evolution of life ending up with advanced simian primates.

evolved through many sophisticated species, to primates, to intelligent primates, and onto humans (Figure 4.1.1).

HUMAN EVOLUTION

As thousands of species evolved including hundreds of species of reptiles, fish, mammals, and aqueous mammals, evolution derived from umpteen types of spontaneous and other forms of mutations. Different size mammals evolved, two-legged Mammals evolved, mammals with larger claws, larger mouth, and larger teeth all evolved. However, never did a large brain species evolve, or a species with a more extensive nervous system evolve. This led paleontologists and archeologists to question how humans ever evolved, or when large brain evolution occurred.

Looking at the evolution of humans from primates (Figure 4.1.2), it appears that five consecutive brain expansion steps were needed for humans to evolve with a large brain. Science shows that Lower Simian primates evolved with a bigger brain (0.17% of body weight) than Prosimian primates, their ancestors (brain size 0.07% of body weight). It appears that Advanced Simian primates (brain 0.74%) evolved with a brain bigger than Lower Simian primates, that Early Hominids evolved with a brain (brain 1.2%) bigger than Advanced Simian primates, and that Homo sapiens evolved with a brain (brain 2.4%) bigger than Early Hominids (Figure 4.1.2). It is like five consecutive larger brain groups of species occurred in the evolution of humans, or that five brain expansion steps occurred consecutively in the evolution of humans (Figure 4.1.2). How did this happen when there was no previous evolution history

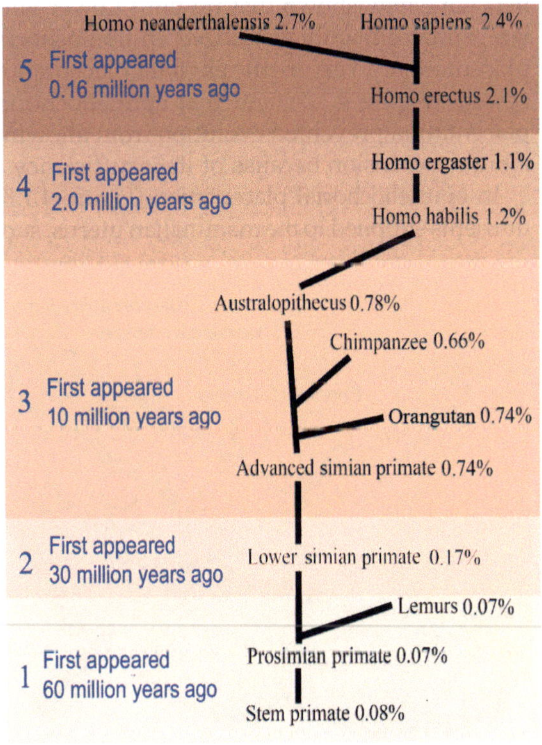

FIGURE 4.1.2 Evolution of the human brain from primates. Percentage value is brain size.

of brain expansion? It is suggested that you read expert remarks on this evolutional nonsense:

- Cunnane SC, Herbige LS, Crawford MA. The importance of energy and nutrient supply in human brain evolution. Nutr Health 1993; 9:19-35.
- Martin RD. Relative brain size and basal metabolic rate in terrestrial vertebrates. Nature 1981; 293:57-60.
- Martin RD. Human reproduction: a comparative background for medical hypotheses. J Reprod Immunol 2003; 59:111-135.
- Leigh SR. Brain growth, life history, and cognition in primate and human evolution. Am. J. Primatology 2004;62:139–164.
- Herculano-Houzel S. The Human Brain in Numbers: A Linearly Scaled-up Primate Brain Front Hum Neurosci 2009;3:31.

Contrary to Darwin, many believe an external factor like God must be the explanation. Scientists have now shown that up until Lower Simian primates mammals utilized epitheliochorial placentation, a very inefficient fetal placentation system (Figure 4.1.3), and that epitheliochorial placentation prevented evolution from allowing a brain expansion because of its insufficiency.

In epitheliochorial placentation (Figure 4.1.3), nutrients supplied to the mammalian uterus, such as glucose and oxygen, had to leave the endometrial artery and cross connective tissue to get to the uterine epithelium, then cross the uterine epithelium, the uterine cavity space, the chorionic epithelial cells, and chorionic connective tissue in order to reach the fetal vascular circulation. This was grossly inefficient and led to the inability of mammal brain development. How was this epitheliochorial placentation problem fixed? How did the fix continue to get consecutively better such that the human brain developed?

hCG EVOLUTION

In this section, the evolution of human chorionic gonadotropin (hCG) and hyperglycosylated hCG, two molecules previously described in Chapter 3.10 are explained. These two molecules drove the evolution of humans. Chorionic gonadotropin (CG, not human so no h) first evolved in Lower Simian primates 30 million years ago from the pituitary gonadotrope cell hormone luteinizing hormone (LH) (Figure 4.1.4). The LH ß-subunit underwent a deletion mutation leading to a frame shift in structure to produce CG ß-subunit (Figure 4.1.5). This led to the development of a 145 amino acid hCG ß-subunit peptide replacing the 121 amino acid LH ß-subunit, or a

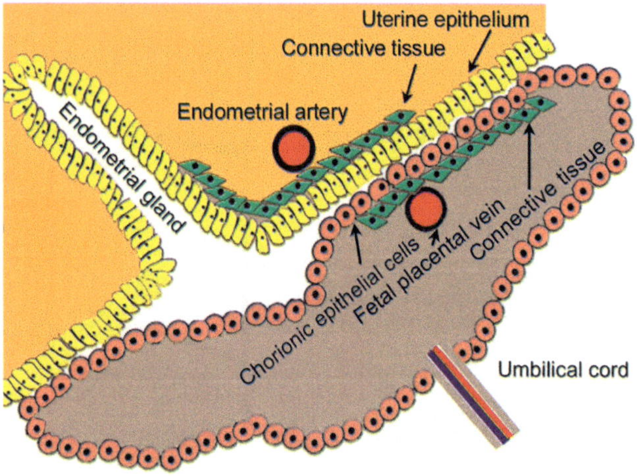

FIGURE 4.1.3 Epitheliochorial placentation.

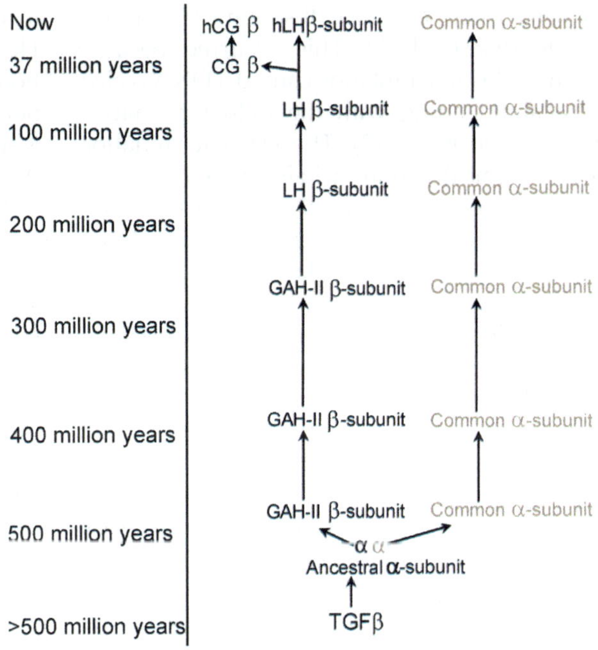

FIGURE 4.1.4 The evolution of hCG from TGFß.

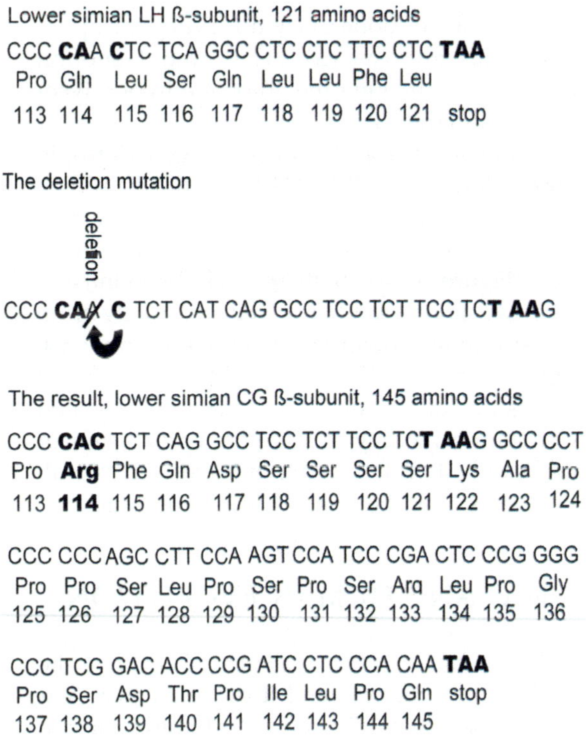

FIGURE 4.1.5 Deletion mutation in LH ß-subunit gene leading to formation of hCGß.

CG with a new extended ß-subunit C-terminal peptide (Figure 4.1.5). This occurred because the single deletion mutation caused a DNA codon frame shift ending with the LHß termination codon (in position 122). The DNA termination codon became the amino acid lysine, and the molecule became extended until a second DNA termination codon was reached at the DNA corresponding to residue 146 (Figure 4.1.5). This extended the molecule from 121 to 145 residues. This combined with the α-subunit, which is common to both molecules, so that both LHß and CGß shared the same α-subunit.

Therefore, unlike pituitary gonadotrope cell LH, multiple forms of CG exist. Placental cytotrophoblast cells are the root cells made fresh. Soon after they are produced, the cells fuse together to make multinuclear syncytiotrophoblast cells, two different cells, cytotrophoblast and syncytiotrophoblast. Both cells by nature had to have the new CG genes. Cytotrophoblast cells glycosylated CG α- and ß-subunit one way making the hormone CG, while syncytiotrophoblast cells glycosylated CG α- and ß-subunit another way making the autocrine hyperglycosylated CG (h-CG).

LH evolved from the molecule gonadotropin ancestral hormone-II (GAH-II) about 280 million year ago (Figure 4.1.4), and GAH-II evolved from an ancestral α-subunit about 500 million years ago (Figure 4.1.4). Both α- and ß-subunits of CG, LH, and GAH-II evolved separately from transforming growth factor-ß (TGF-ß), over 500 million years ago (Figure 4.1.4). The amazing part of this evolution story is that hyperglycosylated CG started binding a TGF-ß receptor again or expressing its TGFß history, but evolved 500 million years after the original TGF-ß molecule.

hCG AND HUMAN EVOLUTION

Fascinatingly, the evolution of CG drove the evolution of humans, and it drove all five clear evolutionary steps (Figure 4.1.2) in brain expansion in the development of primates, hominids, and Homo sapiens. CG first appeared as a result of deletion mutation in the LH ß-subunit gene in Lower Simian primates. Mutation then occurred in hCG ß-subunit and later appeared as a super-CG in Advanced Simian Primates.

Further mutations occurred, and CG next appeared in Early Hominids as a super-duper CG. Finally, CG (or hCG) appeared as ultra-super-duper CG (or hCG) in Advanced Hominid or humans.

Human CG is approximately 100X more potent than its ancestor hormone LH. CG promoted the formation of a new form of fetal placentation, hemochorial placentation, much more efficient than mammalian epitheliochorial placentation. Hemochorial placentation was a system in which the placenta imbedded in the uterus. A maternal blood tank is formed and nutrients only had to cross inside this tank one layer of cells to reach the fetal circulation (Figure 4.1.6). hCG and hyperglycosylated hCG drove most features needed for hemochorial placentation, growth of placental villous tissue (shown by large back and white tree structures in Figure 4.1.6), growth of maternal uterine spiral arteries, and growth of an umbilical circulation within placental villous tissue (Figure 4.1.6).

Primitive hemochorial placentation, which evolved as a consequence of Lower Simian CG, permitted brain development in these primates (Figures 4.1.7 and 4.1.8). Super CG (longer circulating half-life, more potent) permitted super hemochorial placentation in Advanced Simian primates leading to super-brain development. Super-duper CG (even longer circulating half-life, more potent) permitted super-duper hemochorial placentation in Hominids leading to super-duper brain development (Figure 4.1.8). Finally, ultra-super-duper CG (very much longer circulating half-life, more potent) permitted ultra-super-duper hemochorial placentation leading to the development of Homo sapiens.

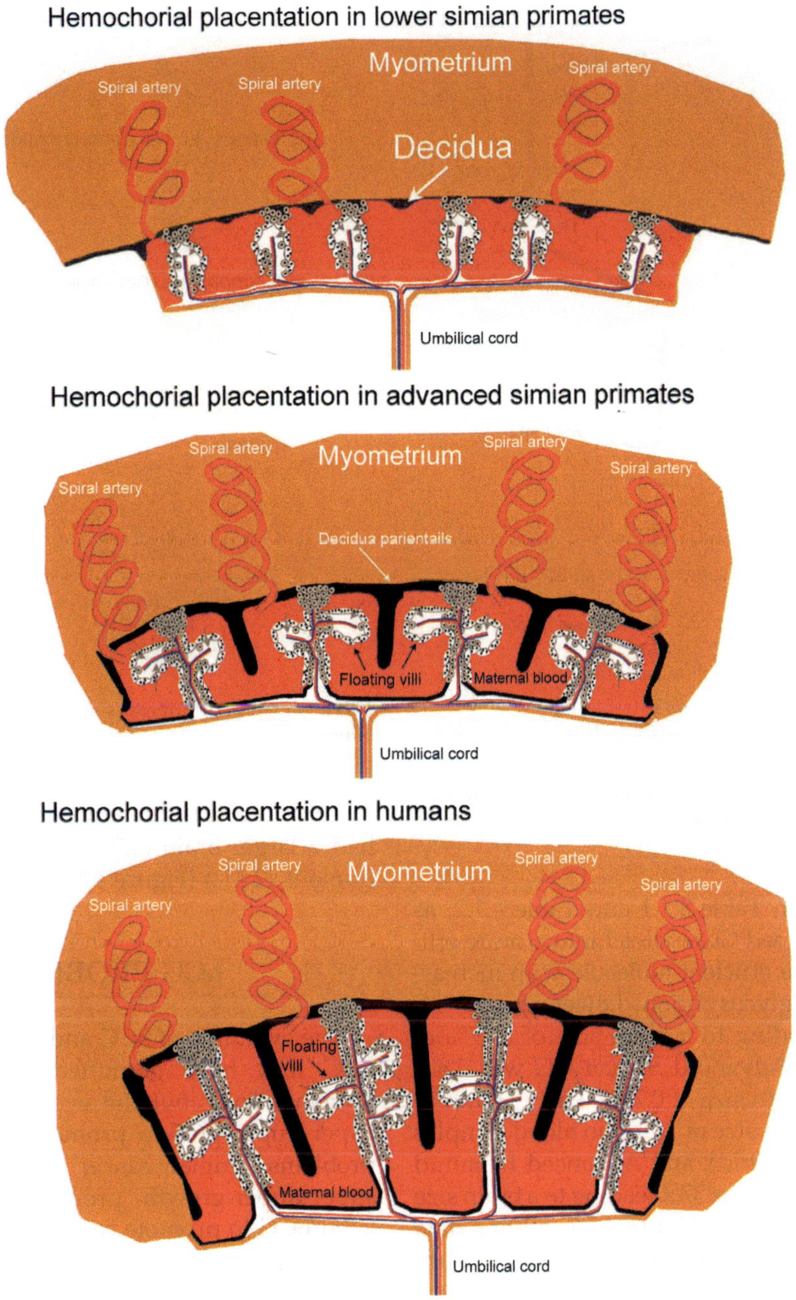

FIGURE 4.1.6 Hemochorial placentation in primates and humans.

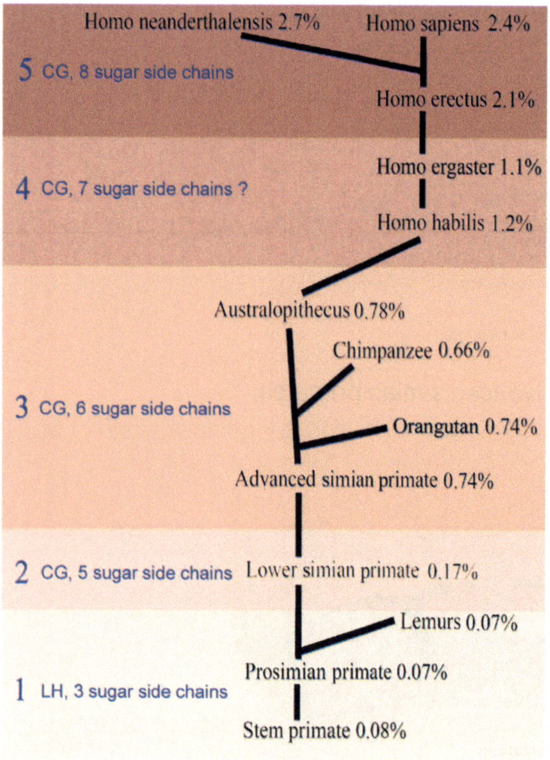

FIGURE 4.1.7 The five-step development of Homo sapiens.? means no data available.

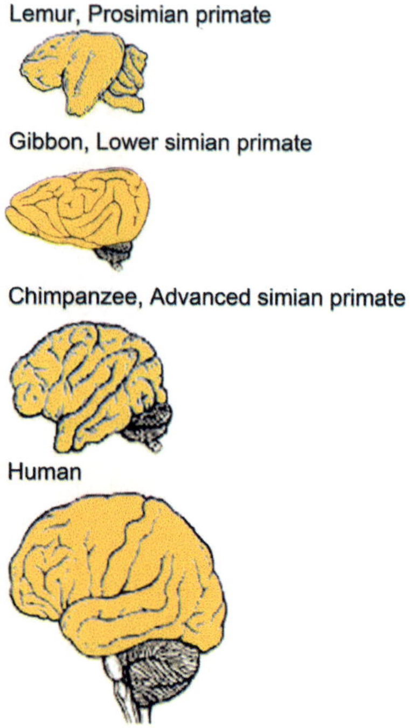

FIGURE 4.1.8 Comparisons of human and primate brains.

As shown in Table 4.1.1 and Table 4.1.2, as species developed CG evolved into a more efficient and more efficient molecule with increasing extent of glycosylation. Lower Simian CG was 0.40 × leading to a brain size of 0.17% of body weight. Advanced Simian CG was 2.4 × leading to a brain size of 0.74%, Early Hominid CG led to a brain size of 1.2% (no blood samples from Early Hominid), and Advanced Hominid CG or human CG was 37 × leading to a brain size of 2.4%. As shown in Figure 4.1.7, all this data fits in with the five-step information on development of Homo sapiens.

Homo neanderthalensis evolved parallel to humans. Science indicates that Homo sapiens either inter-mated with Homo neanderthalensis or killed this species. Homo Neanderthalensis had a larger brain (2.7% of body weight) than humans (2.4%) (Figure 4.1.7).

hCG PROBLEMS

The evolution of hCG and hyperglycosylated hCG may have permitted the evolution of human beings, but the evolution of an ultra-super-duper growth promoter comes with its problems. Human cancer cells steal the ultra-super-duper-growth promoter hyperglycosylated hCG to promote cancer cell super growth and cancer cell super invasion (see Chapter 5.3). It may be amazing that nature developed the ultra-super-duper-growth promoters needed for human brain development, but human cancers abuse these incredible molecules.

TABLE 4.1.1 Properties of CG, Super CG, Super-Duper CG, and Ultra-Super-Duper CG. Isoelectric Point (pI) is a Measure of Acidity. Means Unknown, Early Hominid Blood Unavailable

Species	Structure	Isoelectric point	Circulate ½-life = bioactivity	Brain size % Body weight (%)
Prosimian primate LH	3 sugar side chains	pI=9.0	0.40 h or 0.40×	0.07
Lower simian primate CG	5 sugar side chains	pI=6.3	2.4 h or 2.4×	0.17
Advanced simian primate CG	6 sugar side chain	pI=4.9	6.3 h or 6.3×	0.74
Early hominid CG	7 sugar side chains	?	?	1.2
Advanced hominid CG	8 sugar side chains	pI=3.5	37 h or 37×	2.4

TABLE 4.1.2 Hemochorial Placentation in Primates, Hominids, and Humans. Means no Early Hominid Pregnancy Studied

Species	Invasion depth into uterus % of uterine thickness	Estimated thickness of placenta (from photos and histology slides)
Prosimian primate LH	No invasion	2 mm
Lower simian primate CG	1%	5 mm
Advanced simian primate CG	10%	10 mm
Early hominid CG	?	?
Advanced hominid CG	33%	25 mm

CHAPTER KEYWORDS

Know these keywords and what each word refers to as well as its function. Read over the list. Are all these keywords familiar?

Advanced hominid
Advanced simian primate
Amplification mutation
Ancestral α-subunit
Chorionic epithelial cells
Chromosomal translocation
Darwin
Deletion mutation
Early hominid
Epitheliochorial placentation
Evolution
GAH
Galapagos
Gonadotropin ancestral hormone
Hemochorial placentation
Homo erectus
Homo neanderthalensis
Homo sapiens
LH
Lower simian primate
Luteinizing hormone
Mammals
Mutagens
Point mutation
Prosimian primate
Spontaneous mutation
TGF-ß
Transforming growth factor-ß
Umbilical cord
Uterine epithelium

INTERNET REFERENCES

Use these references to find out more.
en.wikipedia.org/wiki/Human_evolution
humanorigins.si.edu/resources/intro-human-evolution

evolution.berkeley.edu/evosite/evo101/IIE2cHumanevop2.shtml
www.ncbi.nlm.nih.gov/pmc/articles/PMC3351023/

CHAPTER QUIZ

Answer these questions. Do you know the answers, holy cow! Check yourself out in this quiz, and prove that you are a humanology wiz.

1. Which is not a point mutation?
   a. A change in DNA sequence of a single base
   b. Change of a T base for an A base
   c. Change of a C base for a G base
   d. Change of a T base for a G base
   e. Deletion of a DNA base
2. Epitheliochorial placentation was an inefficient placenta system in mammals because:
   a. Nutrients had to cross a thick layer of connective tissues
   b. The chorionic epithelium of mammals was blocked by fatty acid cells
   c. The fetal vein of mammals was not readily penetrable
   d. Nutrients had to cross a uterine cavity, leading to insufficiency
   e. Nutrients had to cross multiple cell layers to reach the fetal circulation
3. Hemochorial placentation involves:
   a. Tanks of maternal blood and only a single layer of cell to reach fetal circulation
   b. Tanks of maternal blood and multiple layers of placental cells to reach the fetal circulation
   c. Fast flowing maternal blood and a single layer of cell to reach fetal circulation
   d. Fast flowing maternal blood and multiple layers of placental cells to reach the fetal circulation
   e. Spiral artery blood and spiral fetal umbilical circulation
4. The evolutionary rise in the efficiency of hCG as a hormone from Lower Simian primates to Advanced Simian primates was due to what molecular changes?
   a. The addition of more acidic Glu and Asp amino acids
   b. The addition of more basic Lys and Arg amino acids
   c. The addition of more sugar side chains
   d. The addition of glucose residues to CG
   e. The addition of amine groups to the hCG ß-subunit peptide
5. Homo neanderthalensis had a
   a. Larger brain than humans
   b. Slightly smaller brain than humans
   c. Much smaller brain than humans
   d. Same size brain than humans
   e. Double-size brain compared to humans

CHAPTER

4.2

Modern Human Pseudo-Evolution

The earliest Homo sapiens that evolved in the wilds of Africa were wild animals that hunted for food and lived in the wild and did not think in any way. Homo sapiens were not the first hominid, and probably learned some symbol and sound communication skills from their evolutionary predecessor Homo heidelbergensis, and from their neighbors Homo neanderthalensis.

Homo sapiens probably learned simple symbols to communicate with other humans. Hanging its hands out in front might have meant "Go Away," sticking its finger in its mouth meant "I'm Hungry," and smiling might have meant that the male human was attracted to a woman, or something like this. The Pooh-pooh psychology theory suggests that the first words might have been emotional interjections and exclamations. The Ding-dong theory of Muller and colleagues suggests that natural vibrating noises may have been the basis of the first spoken words. It hypothetically took almost 50,000 years for symbols to be replaced with simple noises, "Ugh" might have meant "Go Away," "Agh" might have meant "I'm Hungry," and "Yeh" might have been appropriate to show attraction to a female. It was seemingly almost 100,000 years before noises and signs became a simple language, "Go" for "Go Away," "Food" for "I'm Hungry," and "TaTa" for goodbye.

With the development of a simple language and ability to think in that simple language, humans started thinking and civilization developed. Quickly, once humans started thinking in a simple language a complex language, as used today, evolved (Figure 4.2.1). Data presented here is a middling of hypotheses proposed (http://en.wikipedia.org/wiki/Origin_of_language). This is theoretically how humans pseudo-evolved. They evolved because they had a complex brain and were able to conceive the importance of communication. The evolution of language can only be properly traced with a time machine. Hypothetically, this is how it evolved (Figure 4.2.1).

Human skeletons were found in Africa dating up to 195,000 years old, 160,000, 120,000, and 115,000 years old. Skeletons were found in the Middle East and South Africa dating to 100,000 years old. Human remains have been found in Europe and Asia dating to about 40,000-60,000 years ago. Clearly humans evolved in Africa, and eventually migrated to the Middle East and then on to Europe and Asia.

Genetic studies have shown that major variations in the Y chromosome and chromosome 12

Human Physiology, Biochemistry and Basic Medicine
http://dx.doi.org/10.1016/B978-0-12-803699-0.00035-9

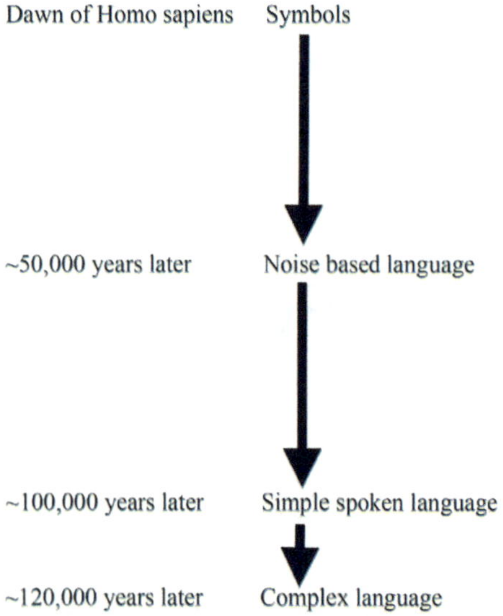

FIGURE 4.2.1 The hypothetical evolution of language in humans. Data is a middling or summary of expert theories (Bow-wow, Pooh-pooh, Ding-dong, Yo-he-ho, and Ta-ta hypothesis) regarding the evolution of language (http://en.wikipedia.org/wiki/Origin_of_language).

have occurred in the modern evolution of humans. Are we genetically different than our human ancestors we ask? Yes, we clearly are. It appears that our species evolved after the end of the last ice age, 10,000 year ago. In this respect, we are seemingly a further-evolved Homo sapiens species or Homo superiorensis.

The Middle Paleolithic period, 40,000-200,000 years ago, was the time during which archaic humans, Homo sapiens, and Homo neanderthalensis first appeared and flourished. These peoples used stone hand axes as tools to survive. The living method in the Middle Paleolithic period included scavenging, but there is also clear evidence of hunting and gathering activities. Deliberate human burials, with evidence of ritual behavior, are also found during this period.

55,000 years ago, archaic humans tended to their elderly. Some evidence for cannibalism is also found at this time. The Middle Paleolithic ends with the gradual disappearance of the Neanderthal for uncertain reasons and the ascendancy of Homo sapiens about 40,000-55,000 years ago.

The Upper Paleolithic Age or next age, 40,000-10,000 years ago, was a period of transition (Figure 4.2.3). The Neanderthals edged out and disappeared and modern humans began to have the world to themselves. Stone tools of the Upper Paleolithic Age were blade-based. Blades were at that time stone pieces twice as long as they are wide, with parallel sides, and they were used to create an astonishing range of tools with specific purposes. In addition, bone, antler, shell, and wood were used for both artistic and working tools, including the first eyed needles presumably for making clothing (about 21,000 years ago).

The Upper Paleolithic Age is perhaps best known for cave art, wall paintings, and engravings of animals and abstractions at caves (Figure 4.2.3). People living during the Upper Paleolithic lived in houses with dugout floors, fireplaces, and windbreaks. Hunting became more focused, more seasonal, and more sophisticated, as shown by the remains of animals and selective butchery. Food storage was practiced. Evidence suggests that small groups of people went on hunting trips and returned with meat to the base camps. The end of the Upper Paleolithic Age came about because of climate changes and global warming.

The Neolithic Revolution is sometimes called the Agricultural Revolution (Figure 4.2.3). It was the wide-scale transition of human cultures from a lifestyle of hunting and gathering to one of agriculture and settlement. Archaeological data indicates that the farming of various types of plants and animals evolved in many locations, starting around 12,000 years ago (Figure 4.2.2).

The Neolithic Revolution involved far more than the adoption of a limited set of food-producing techniques. It transformed the small groups of hunter-gatherers that had dominated pre-history into societies based in villages and towns. These societies modified their environment by means of specialized food growth and employed food storage methods, allowing

FIGURE 4.2.2 Neolithic pottery.

Middle Paleolithic period, 40,000-200,000 years ago
Homo sapiens and Homo neanderthalensis both flourished
They had stone axes
Hunting
Human burials
Ritual experiences

Upper Paleolithic period, 10,000-40,000 years ago
Homo neaderthalensis became extinct
Cave art
Lived in houses
Seasonal group hunting and butchery
Food storage
Making clothing

Neolithic period, 3000-12,000 years ago
Farming and agricultural revolution
Farming of vegetables and animals
Community gatherings
Living as a communots in villages and towns
Trading
Centralization of administration
Division of labour

Civilization period, present-3000 years ago
Centralization of government and administration
Formation of cities
Community life
Community food supply
Social class structure
Organization of religions

FIGURE 4.2.3 Development of humans.

surplus food production. These developments provided the basis for densely populated settlements, which led to division of labor, trading economies, centralized administrations, property ownership, and fights over fertile properties.

Civilizations were societies organized into populated settlements, divided by social classes with ruling elite and subordinate populations. The earliest emergence of civilizations is associated with the final stages of the Neolithic Revolution. The Neolithic revolution was dependent upon the development of grains, domestication of animals, and the development lifestyles that allowed the development of economies and the accumulation of surplus products.

Toward the end of the Neolithic period, various civilizations began to rise, starting around 3300 BCE with the development of Egyptian society. The Iron Age around 1200 BCE saw many new societies emerge. Each society carefully examined basic sciences and the principles of living and law leading to what we today call civilization. Religion and communal religious practice is clearly part of civilization. The Jewish beliefs in God and communal religious practices led to civilization, as did the later Christian faiths and Moslem faiths. Indeed, the Egyptian polygamous faiths and Roman polygamous faiths aided the development of civilized societies, even though the polygamous religious beliefs were eventually abandoned (Figure 4.2.3).

CHAPTER KEYWORDS

Know these keywords and what each word refers to as well as its function. Read over the list. Are all these keywords familiar?

Agricultural revolution
Civilization
Homo neanderthalensis
Homo sapiens

Iron age
Middle Paleolithic
Neolithic
Neolithic revolution
Upper Paleolithic

INTERNET REFERENCES

Use these references to find out more.
en.wikipedia.org/wiki/Civilization
en.wikipedia.org/wiki/Neolithic
en.wikipedia.org/wiki/Neolithic_Revolution
teachmiddleeast.lib.uchicago.edu/foundations/origins-of-civilization/essay/essay-02.html
history-world.org/neolithic.htm
en.wikipedia.org/wiki/Middle_Paleolithic
archaeology.about.com/od/mterms/qt/middle_paleolit.htm
humanorigins.si.edu/evidence/behavior/tools/middle-tools
http://en.wikipedia.org/wiki/Neanderthal
http://en.wikipedia.org/wiki/Origin_of_language

CHAPTER QUIZ

Answer these questions. Do you know the answers, holy cow! Check yourself out in this quiz, and prove that you are a humanology wiz.

1. Homo sapiens first evolved in which age?
   a. Neolithic
   b. Neolithic revolution
   c. Upper Paleolithic
   d. Middle Paleolithic
   e. Iron Age
2. Which fact is true about human language?
   a. Humans spoke with each other shortly after their evolution.
   b. The first humans spoke Afrikaans.
   c. It took humans about 100,000 years to develop an oral language.
   d. It took humans about 10,000 years to develop an oral language.
   e. It took humans about 1000 years to develop an oral language.
3. Which fact is true?
   a. Farming seemingly started in the Neolithic era.
   b. Early farming was limited to one-acre sites.
   c. Farming seemingly started in the Upper Paleolithic era.
   d. Farming seemingly started in the Middle Paleolithic era.
   e. Farming seemingly started in the Iron Age.

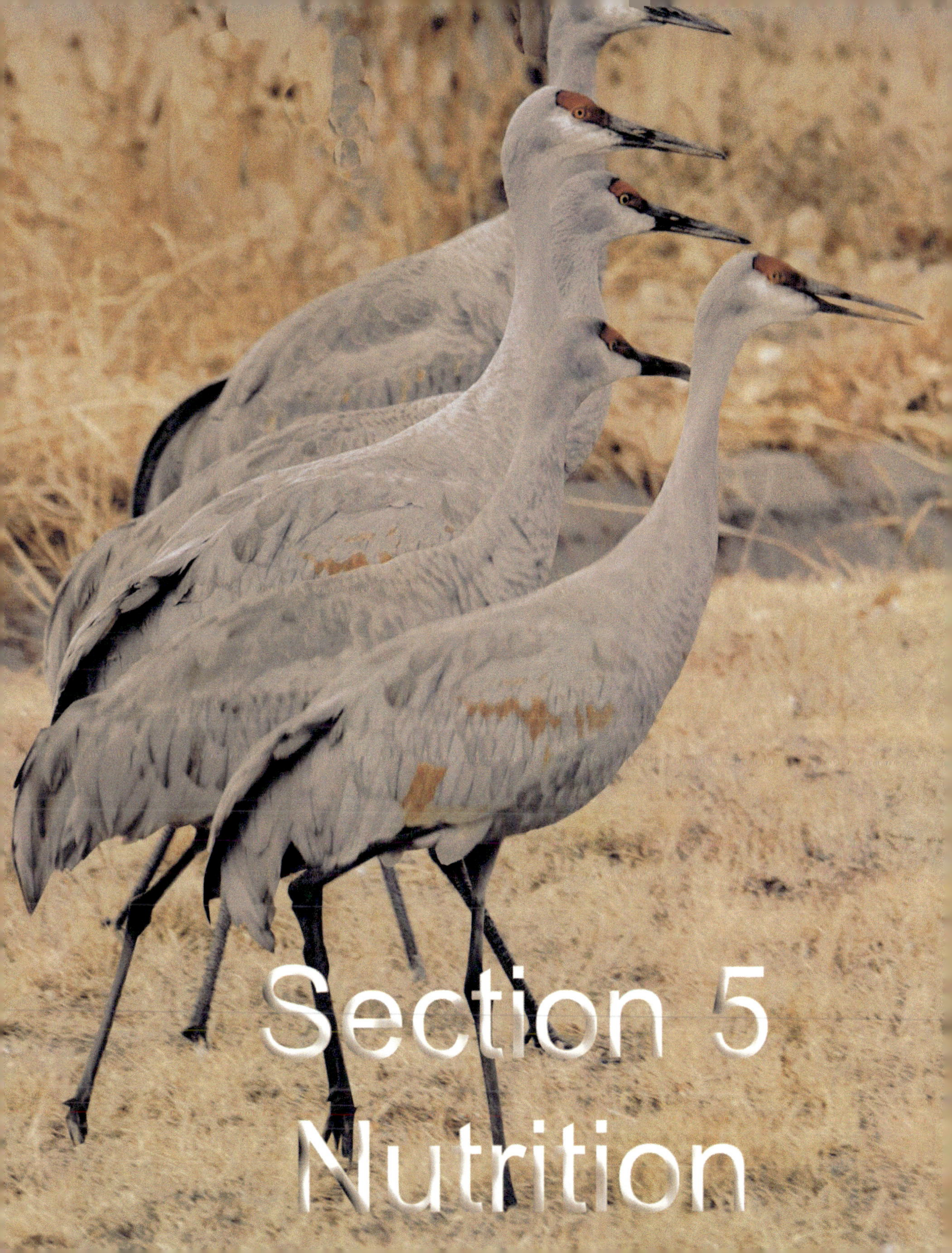

Section 5
Nutrition

CHAPTER

5.1

Macronutrients

LIPIDS AND FATS

The major nutrients in the human diet consist of fats, carbohydrates, and proteins. These biochemicals provide for most of the structural and energy requirements of the human organism.

Fats and oils belong to the water-insoluble class of compounds known as lipids. In addition, lipids also include waxes, sterols, fat-soluble vitamins, eicosanoids, terpenes, and other compounds. Chemically fats are triacylglycerols (also called triglycerides), i.e., fat molecules which are comprised of three fatty acid moieties linked to glycerol through ester linkages. Fatty acids are chemically classified as carboxylic acids and are distinguished by the presence of a carboxyl group. Oils differ from fats in that they are liquids at room temperature. Animal fats are about 40% saturated and 60% unsaturated. Unsaturated fats differ from saturated fats by the presence of double bonds in their fatty acids. The fatty acids of unsaturated fats can be monounsaturated, wherein the fatty acid has a single double bond, or they may be polyunsaturated, containing more than one double bond. Most fatty acids have carbon chains containing from 12 to 20 carbon atoms and have an even number of carbons.

Chemically, the configuration of double bonds in unsaturated fatty acids can be *cis* or *trans*. Double bonds formed in fatty acids from human metabolic enzymes have the *cis* configuration. Most naturally occurring unsaturated fatty acids in nature contain bonds with *cis* configurations. One effect of *cis* configured double bonds in fatty acids is to introduce a bend in the structure. The departure from linearity makes it more difficult for *cis*-unsaturated fatty acids to interact with other fatty acid molecules through dispersion forces, thereby decreasing the melting point. Fats containing *cis*-fatty acids or containing shorter chain saturated fatty acids also tend to have lower melting points tending to be liquids at room temperature. Because *trans* fats can more efficiently pack together in the solid state, they more readily form plaques on artery walls. In addition, human cells lack the metabolic enzymatic machinery to process *trans* fats effectively so that their circulating half-lives are extended, thus further contributing to arterial plaque formation.

Human Physiology, Biochemistry and Basic Medicine
http://dx.doi.org/10.1016/B978-0-12-803699-0.00005-0

NOMENCLATURE AND REPRESENTATION OF FATTY ACIDS

Because fatty acid molecules can be quite long, they are usually represented by drawing only a skeletal structure showing carbon-carbon bonds and carboxyl oxygens. Skeletal structures do not show hydrogen atoms or the carbon-hydrogen bonds. It is important to remember that carbon atoms have four bonds. A carbon atom shown with three bonds shown in a skeletal structure also has a fourth bond to a hydrogen atom. Dodecanoic acid, a 12-carbon fatty acid, is shown in Figure 5.1.1 as a condensed structural formula and skeletal formula.

Fatty acids are named with common names and with systematic names according to the scheme devised by the National Union of Pure and Applied Chemistry (IUPAC). According to systematic naming, the carboxyl carbon is assigned as C1, with numbering ascending along the chain. According to common nomenclature, the carbon atoms after the carboxyl carbon are designated with sequential Greek characters α, β, γ, δ, ε, etc. Thus, for example, IUPAC C4 of a fatty acid corresponds to the γ carbon according to common nomenclature. The carbon farthest from the carboxyl carbon is often designated as ω or n. Biochemists frequently use a shorthand notation when referring to fatty acids in which the number of carbon atoms is listed first followed by a colon and the number of double bonds in the fatty acid and then a list of the positions and configurations of the double bonds. Configurations of double bonds can be either *cis* or *trans* and are indicated with c or t, respectively. Most double bonds in unsaturated fatty acids naturally occur with *cis* configuration so that fatty acids containing only *cis* double bonds can be written, omitting *cis* designation.

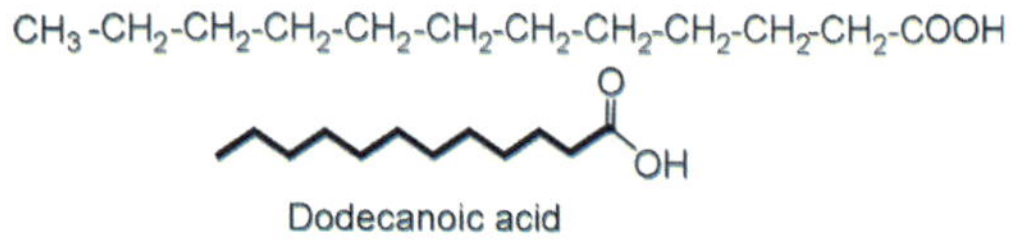

FIGURE 5.1.1 Two structural formulas depicting dodecanoic acid.

One function of fat is to store energy from excess calories in the diet. Fats contain about 9 kcal/g.

ESSENTIAL FATTY ACIDS

Alpha-linolenic acid (ALA), all-*cis* 18:3($n-3$) and linoleic acid, all-*cis* 18:2($n-6$) are the two essential fatty acids required in the diet for good health but that cannot be synthesized in the body. The essential fatty acids are precursors of lipids with diverse physiological functions. These fatty acids contain multiple double bonds and are members of the class of polyunsaturated fatty acids (PUFAs).

Linoleic acid is an essential fatty acid and a precursor of eicosanoids that is produced from a common intermediate, arachidonic acid 20:4 ($n-6$). An important class of lipids involved in inflammation, the prostaglandins, is synthesized from arachidonic acid. Pathways for synthesis of eicosanoids and other compounds are illustrated at the Kyoto Encyclopedia of Genes and Genomes (KEGG) database. Natural sources of linoleic acid include safflower oil, soybean oil, sunflower oil, and corn oil. Thromboxanes, which are also eicosanoids, are involved in inflammation. The enzyme prostaglandin-endoperoxide synthase (also called cyclooxygenase or COX) is a key enzyme in the synthesis of prostaglandins and thromboxanes. The anti-inflammatory effect of aspirin is largely due to its inhibition of COX.

ALA is a precursor of some components of fish oil, eicosapentaenoic acid (EPA; 20:5, $n-3$), and docosahexaenoic acid (DHA; 22:6, $n-3$). Sources of ALA include flax seed and nuts (Figure 5.1.2).

EPA is an $n-3$ PUFA derived from fish oil. It is a precursor of prostaglandin E_3 and some other eicosanoids. EPA seems to display

FIGURE 5.1.2 Structure of α-linolenic acid.

beneficial effects for some human disorders associated with inflammation. Inflammation-resolving activity appears to be due to metabolites of EPA and eicosanoids such as lipoxins, resolvins, and protectins. The anti-inflammatory effects of the n-3 PUFAs act in opposition to the proinflammatory effects associated with many n-6 PUFAs.

The fats from marine animals have greater quantities of longer carbon chain fatty acids and with higher levels of unsaturation when compared to mammalian fats. Increases in dietary n-3 FAs reduce the n-6 fatty acid content in cell membranes, thereby lowering available levels of arachidonic acid and eicosanoids derived from AA such as thromboxane A_2, a stimulator of platelet aggregation and vasoconstriction, and prostaglandin E_2. Prostacyclin, PGI_2, is also produced from AA, via COX. Prostacyclin has antiaggregatory and vasodilator effects, actions that oppose TXA2 vasoconstriciton and platelet aggregation.

PROTEINS

Proteins comprise a major component of human cells having structural and functional roles. Proteins in the human body are constructed from 20 α-amino acids, though some modified amino acids do occur in human proteins. The α-amino group of an amino acid is located on carbon 2 of the amino acid, which is often designated as the α-carbon. Except for the simplest amino acid, glycine, the α-amino acids exhibit optical isomerism and can exist in the D or L configuration. The naturally occurring amino acids in humans have the l configuration. From a nutritional perspective, proteins can be categorized as essential or nonessential.

TABLE 5.1.1 Essential Amino acids

Essential amino acid	3-Letter abbreviation	1-Letter abbreviation
Phenylalanine	Phe	F
Histidine	His	H
Isoleucine	Ile	I
Leucine	Leu	L
Lysine	Lys	K
Methionine	Met	M
Valine	Val	V
Threonine	Thr	T
Tryptophan	Trp	W

Nonessential amino acids can be synthesized within the body and essential amino acids must be supplied in the diet. The nine amino acids generally considered to be essential are listed in Table 5.1.1.

Because proteins are synthesized from constituent amino acids that are linked together sequentially, all of the amino acid residues required for the polypeptide chain of a protein to be synthesized must be available at the time of synthesis. High-quality dietary proteins contain all of the essential amino acids in sufficient proportion to support efficient protein synthesis in the body. Low-quality protein sources may contain all nine essential amino acids but not in the proportions necessary to fulfill the body's needs. A diet that is deficient in any of the essential amino acids will result in negative nitrogen balance, a condition in which more protein is degraded than is synthesized as indicated by excretion of nitrogen derived from proteins. Protein quality is often expressed as biological value (BV) by comparison of amino acid composition to a high-quality reference protein such as egg. The reference protein is typically assigned a value of 100. Assuming sufficient caloric intake, protein with BV of 70 should be adequate to support growth. Other measures are also used to compare the values of dietary proteins. Net protein utilization (NPU) includes both BV and digestibility of proteins and is a measure of dietary nitrogen retention. Protein digestibility can be lower when the protein source is vegetable matter due to nonprotein constituents that interfere with digestion. The protein efficiency ratio (PER) is also used to compare dietary protein utilization. This method is based on measurement of the weight gain of a test animal in relation to protein intake. The pathways for essential amino acids tend to be much more complex than the pathways for synthesis of the nonessential amino acids. This is one of the reasons an organism such as rice, which must synthesize all of its amino acids, has more genes in its genome than humans. It is a matter of greater metabolic overhead. By relying on other organisms to synthesize the essential amino acids, humans and other animals higher on the food chain reduce their metabolic overhead and produce proteins with functions not found in plants, such as those that function in vision and neurotransmission.

If a dietary protein source is deficient in one of the essential amino acids, one strategy to improve the diet could be simply to supplement the diet with the deficient amino acid. This approach is used in feeding lysine to livestock. Another way to improve the diet with low-quality protein sources is to combine low-quality proteins that complement each other so that the combination is not deficient in any of the essential amino acids. Both bean and corn lack all essential amino acids in sufficient proportion, but the combination of bean plus corn can provide a high-quality combination of amino acids. Bean plus rice at a ratio of about 6.5:3.5 by weight can also produce a high-quality dietary mix of amino acids. Some other complementary sources are yeast plus grain (such as wheat flour and yeast in bread) and perhaps the most famous combination, legume plus grain (peanut butter on bread, optionally with jelly).

TABLE 5.1.2 Recommended Dietary Allowance for Protein

	Grams of protein needed each day
Children aged 1-3	13
Children aged 4-8	19
Children aged 9-13	34
Girls aged 14-18	46
Boys aged 14-18	52
Women aged 19-70+	46
Men aged 19-70+	56

It is also necessary to consume enough protein to fulfill the body's metabolic requirements. The U.S. Centers for Disease control have published protein intake guidelines (Table 5.1.2).

CARBOHYDRATES

Carbohydrates are composed of carbon, hydrogen, and oxygen and have a general formula that approximates CH_2O. They are polyhydroxy aldehydes or ketones or form polyhydroxy aldehydes or ketones when hydrolyzed. Carbohydrates occur as monosaccharides, disaccharides, oligosaccharides, and polysaccharides. The simplest sugars, mono- and disaccharides, are composed of one or two monosaccharide units. Oligosaccharides are typically classified as carbohydrates comprising 3-10 monosaccharide units. Polysaccharides contain more than 10 monosaccharide units and can be quite large. Complex carbohydrates can be broken down to form simple sugars by hydrolysis.

MONOSACCHARIDES

Monosaccharides occur with carbon chains from 3 to 7 in length (Table 5.1.3). Monosaccharides with aldehyde functionality are termed

TABLE 5.1.3 Monosaccharides

	Trioses	Tetroses	Pentoses	Hexoses	Heptoses
Carbon atoms	3	4	5	6	7

aldoses and those with ketone functionality are termed ketoses. For example, a 5-carbon aldehyde-type monosaccharide, an aldopentose, is both an aldose and a pentose. Glucose is an aldohexose while fructose is a ketohexose. The carbons of monosaccharides are numbered so that the carbonyl carbon has the lowest possible number. Carbohydrates and many other organic compounds such as amino acids exhibit stereoisomerism and optical isomerism. Stereoisomers, though having the same formula and order of chemical groups, differ in the three-dimensional configurations. Optical isomers or enantiomers are stereoisomers which occur as nonsuperimposeable mirror images. Stereoisomers that are not enantiomeric are termed diastereomers. D-Glucose and D-mannose are related as diastereomers, while D-glyceraldehyde and L-glyceraldehyde are enantiomers. D-Glucose, D-mannose, D-glyceraldehyde, and L-glyceraldehyde are shown below as Fischer projections (Figure 5.1.3).

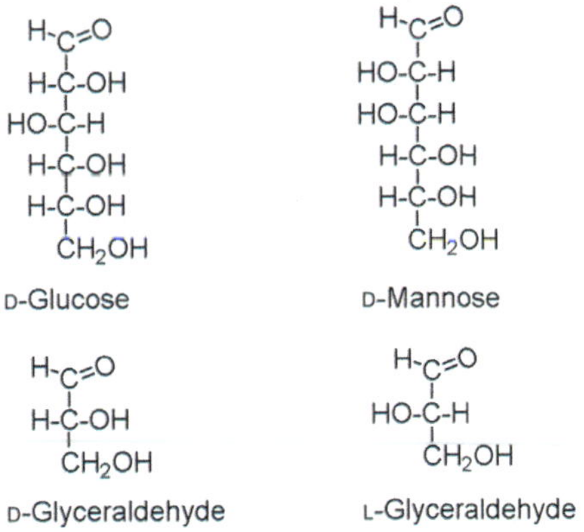

FIGURE 5.1.3 D-Glucose, D-mannose, D-glyceraldehyde, and L-glyceraldehyde.

Monosaccharides with five or greater carbons can have both straight chain forms and cyclic ring forms. D-Glucose can form α-D-glucopyranose or β-D-glucopyranose configurations. These are formed when the oxygen of the hydroxyl group of glucose's carbon 5 adds to the carbonyl carbon to form a hemiacetal. Carbon 6 of glucose is not part of the six-membered pyranose ring (Figure 5.1.4).

Five-carbon monosaccharides such as ribose form five-membered ribofuranose rings (Figure 5.1.5).

D-Fructose, a ketohexose forms five membered fructofuranose rings. D-Fructose is shown below in straight chain and α-D-furanose forms. The ring forms when the O of the C5 hydroxy group forms a bond with C2, the carbonyl

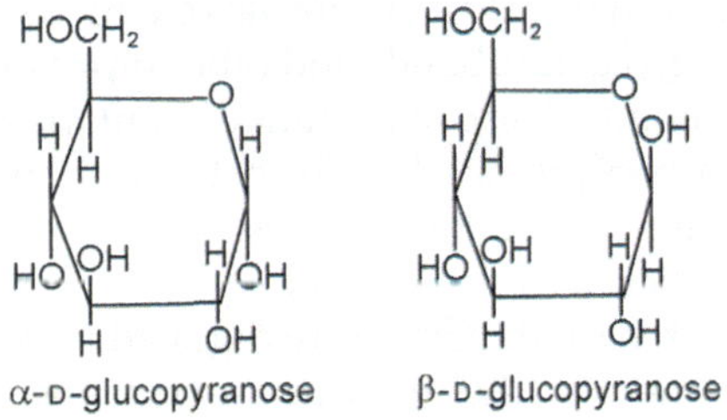

FIGURE 5.1.4 α and β structures of glucose.

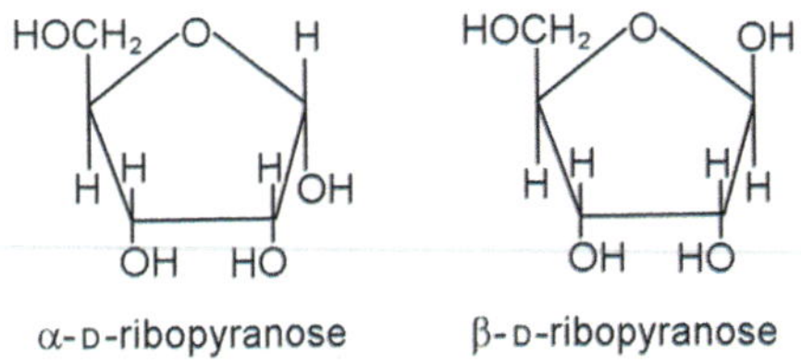

FIGURE 5.1.5 α and β structures of ribose.

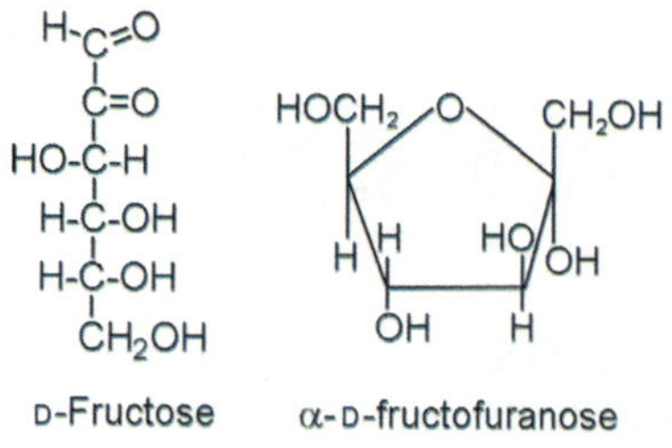

FIGURE 5.1.6 D-Fructose.

carbon resulting in a cyclic hemiketal. Carbons 1 and 5 are not part of the five-membered ring (Figure 5.1.6).

DISACCHARIDES

Disaccharides are composed of two monosaccharide units linked together by a glycosidic bond. The most common glycosidic bonds connecting monosaccharide units are *O*-glycosidic bonds in which the oxygen from a hydroxyl group becomes linked to the carbonyl carbon. Sugars may also become linked to molecules by *N*-glycosidic bonds and other types of glycosidic bonds. The configuration about the anomeric carbon can be alpha or beta as discussed in Chapter 1.3. Some common disaccharides are listed below (Table 5.1.4).

Sucrose and maltose are digested in the small intestine by the enzyme sucrase-isomaltase, which is composed of two subunits, one that hydrolyzes sucrose and maltose and the other that hydrolyzes isomaltose. Maltose is a product of the breakdown of starch by amylase. Lactose is digested in human infants, children, and many adults by lactase. The quantity of lactase secreted in many people may diminish with age. Symptoms of intolerance include gas, belly pain, and bloating. Lactose-free milk is leading to lactose intolerance because of the population's inability to digest lactose. Symptoms of lactose commercially available as are lactase supplements that can be taken with milk or lactose-containing dairy products.

OLIGOSACCHARIDES

Indigestible oligosaccharides from plants include oligosaccharides from plants that are divided into two classes, fructooligosaccharides and galactooligosaccharides. Maltooligosaccharides are hydrolyzed to glucose in the small intestine.

POLYSACCHARIDES

Polysaccharides are polymers of sugar units that occur as storage polysaccharides, such as starch, and structural polysaccharides, such as cellulose. Polysaccharide chains may be linear or highly branched. Starch, which is a major source of caloric energy for most humans, comprises amylose and amylopectin, both of which are made from glucose monomers.

TABLE 5.1.4 Disaccharides

Disaccharide	Common name	Configuration	Monosaccharides
Sucrose	Table sugar	$\beta(1 \rightarrow 2)$	Glucose-fructose
Lactose	Milk sugar	$\beta(1 \rightarrow 4)$	Galactose-glucose
Maltose	Malt sugar	$\alpha(1 \rightarrow 4)$	Glucose-glucose
Isomaltose		$\alpha(1 \rightarrow 6)$	Glucose-glucose

Amylose is composed of glucose linked α(1→4) and has a tightly coiled helical structure. In addition to α(1→4) glycosidic bonds, amylopectin contains α(1→6) linkages with spacing of about 24-30 to yield a highly branched structure. The size of amylose and amylopectin molecules can be quite large. For example, rice amylose has been reported to have a molecular weight of about 3 million Daltons (~18,400 glucose units) and rice amylopectin about 40-50 million Daltons (~246,000 to ~369,000 glucose units). Dietary starch is digested to maltose by α-amylase and is then further hydrolyzed to glucose. Much of dietary fiber is in the form of cellulose. Cellulose is a linear unbranched polymer of D-glucose monomers having β(1→4) linkages. Cellulose is a major structural component of cell walls in green plants. Wood also comprises a substantial amount of cellulose. Cellulose is an insoluble form of dietary fiber. Hemicellulose is another plant cell wall polymer that is indigestible. Hemicellulose is digested by hemicullases from bacteria in the human gut. Hemicellulose comprises several different monosaccharide monomers and is therefore a heteropolymer or copolymer unlike cellulose, which is a homopolymer. Hemicellulose is a soluble fiber and is the major component of soluble fiber in apple. Oat and banana are also good sources of soluble fiber.

CHAPTER KEYWORDS

Know these keywords and what each word refers to as well as its function. Read over the list. Are all these keywords familiar?

α-D-Glucopyranose
β-D-Glucopyranose
Aldohexose
Aldopentose
Amylopectin
Amylose
Anomeric
Arachidonic acid
Carbon-carbon
Carboxyl group
Carboxyl oxygens
Carboxylic acids
cis
Cyclic hemiketal
Cyclooxygenase
Diastereomers
Disaccharides
Disaccharides
Docosahexaenoic acid
Dodecanoic
Eicosanoids
Eicosapentaenoic acid
Essential fatty acids
Ester linkages
Fat-soluble vitamins
Fischer projections
Glucose
Glycosidic bond
Hemiacetal
Hemicellulose
Heptoses
Hexoses
Isomaltose
Ketones
Lactose
Lactose intolerance
Linolenic acid
Lipoxins
Maltose
Monomers
Monosaccharides
Net protein utilization
Oligosaccharides
Oligosaccharides
Optical isomerism
Pentoses
Polyhydroxy aldehydes
Polyhydroxy aldehydes
Polysaccharides
Polysaccharides
Prostaglandin
Prostaglandin-endoperoxide synthase

Protectin
Protein efficiency ratio
Resolvins
Ribofuranose rings
Stereoisomerism
Sterols
Sucrose
Terpenes
Tetroses
Thromboxanes
Thromboxanes
trans
Triacylglycerols
Triglycerides
Trioses
α-Carbon

INTERNET REFERENCES

Use these references to find out more.
en.wikipedia.org/wiki/Protein
en.wikipedia.org/wiki/Fat
en.wikipedia.org/wiki/Carbohydrate

CHAPTER QUIZ

Answer these questions. Do you know the answers, holy cow! Check yourself out in this quiz, and prove that you are a humanology wiz.

1. *cis* fats differ from *trans* fats in that
   - **a.** *cis* fats tend to have lower melting points
   - **b.** *cis* fats are produced in mammals
   - **c.** *trans* fats in the diet may contribute to atherosclerosis
   - **d.** *trans* fats cannot be oxidized
   - **e.** a to c are correct
2. Essential amino acids
   - **a.** Are essential as cofactors for enzymes
   - **b.** Are not synthesized in the human body at sufficient quantity to support growth
   - **c.** Do not need not be supplied from the diet
   - **d.** Are d-amino acids as opposed to l-amino acids
   - **e.** Are neurotransmitters and are essential for neurological function
3. Eicosapentaenoic acid is
   - **a.** An essential fatty acid from fish oil
   - **b.** Formed in fatty tissue in small quantities in humans
   - **c.** A polyunsaturated fatty acid
   - **d.** a and c are correct
   - **e.** a and b are correct
4. When the oxygen of the hydroxyl group of an aldohexose's carbon 5 adds to the carbonyl carbon
   - **a.** A carboxylic acid is formed
   - **b.** A toxic substance is formed
   - **c.** An amino acid is formed
   - **d.** A hemiacetal is formed
   - **e.** A ketone is formed

CHAPTER

5.2

Vitamins and Minerals

Vitamins and many minerals (including many trace elements) are generally required in small quantities as nutrients for good health. One exception is calcium, which is required in greater than trace quantities in the human diet mainly because it is structural component in bones and teeth.

Vitamins are essential organic molecules that cannot be synthesized by the body. However, not all essential organic nutrients are classified as vitamins, such as essential fatty acids and essential amino acids. The vitamins are subdivided into fat-soluble and water-soluble categories. Water-soluble vitamins, after absorption from dietary foodstuffs, are eliminated much more quickly than fat-soluble vitamins, which can be retained in fatty tissues. For this reason, water-soluble vitamins should be consumed daily in the diet while fat-soluble vitamins should be consumed regularly but not necessarily on a daily basis. Most vitamins function as components of cofactors or coenzymes for enzymes and are required for many enzymatic reactions to occur in human metabolism. Many metal ions in minerals also function as cofactors in enzymatic reactions. Because vitamins and other cofactors bind tightly to enzymes, even at low concentrations, they are only required in sufficient quantity to interact with their associated enzymes. Vitamins and minerals in excess of the recommended daily allowances would not be expected to provide any advantage because once the enzymatic reactions have all the cofactors they need, any excess free cofactors will not perform any useful function. This is much like an automobile, which needs only four tires. Additional tires won't have any appreciable effect. In addition to acting as coenzymes, some vitamins also act as hormones (vitamin D) and antioxidants (vitamins C and E). The major vitamins and minerals are listed in Tables 5.2.1 and 5.2.2.

FAT-SOLUBLE VITAMINS A, D, E, AND K

Unlike the water-soluble vitamins, which are eliminated relatively quickly, the fat-soluble vitamins do not need to be replenished on a daily basis because they can accumulate in fatty tissues. Because fat-soluble vitamins can accumulate in the body, it is possible to overdose (hypervitaminosis) on these vitamins. Fat-soluble vitamins like dietary fats require bile salts to be

Human Physiology, Biochemistry and Basic Medicine
http://dx.doi.org/10.1016/B978-0-12-803699-0.00037-2

TABLE 5.2.1 Vitamins

Vitamin	Synonyms	Sources	Deficiency	Metabolic Role
Vitamin C	Ascorbic acid Ascorbate	Citrus fruits	Scurvy	Involved in hydroxylation of proline, antioxidant
Thiamin	B_1	Meats, esp. pork, salmon, legumes, nuts, and grains	Beriberi, Korsakoff's syndrome, optic neuropathy	Component of coenzymes including pyruvate dehydrogenase, α-ketoacid dehydrogenase, and transketolase
Riboflavin	B_2	Milk, cheese, green leafy vegetables, nuts	Cheilosis and angular Stomatitis, dermatitis	Component of enzyme cofactors in many redox reactions
Pantothenic acid	B_5	Whole-grain cereals, legumes, eggs, meat, yogurt	Hypertension	Component of Coenzyme A and phosphopantetheine; prosthetic group for many enzymes
Pyridoxine	Pyridoxamine, pyridoxal, B_6	Meat, fish, poultry, organ meats, soy products, nuts, lentils	Depression, confusion, convulsions	Involved in hemoglobin synthesis, numerous enzymatic reactions involving Schiff base intermediates
Biotin	B_7	Yeast, liver and kidney. Egg yolk, soybeans, nuts and cereals	Biotin deficiency	Coenzyme for carboxylase enzymes
Folic acid	Folate, folinic acid, folacin, vitamin B_9	Broccoli, brussels sprouts, liver, spinach, asparagus, peas, chickpeas, brown rice	Anemia, glossitis, mental symptoms, neural tube defects	Involved in enzymatic transfer of 1 carbon units
Cyanocobalamin	B_{12}, hydroxycobalamin, methylcobalamin	Meat, salmon, cod, milk, cheese, eggs	Pernicious anemia, methylmalonic acidemia	Required for methylmalonyl-CoA mutase and methionine synthase reactions
Niacin	Niacinamide, B_3	Yeast, meat, fish, milk, eggs, green vegetables, cereal, beans	Pellagra	Component of NAD and NADP required for oxidation-reduction reactions
Vitamin A	Retinol, retinal, retinoic acid	Cheese, eggs, yogurt	Night blindness, damage to cornea, GI, and respiratory tracts	Required for formation rhodopsin; hormone-like activity in growth and reproduction
Vitamin D	Cholecalciferol, vitamins D2 and D3, 1, 25-dihydroxy vitamin D	Oily fish, such as salmon, sardines and mackerel, eggs	Rickets, skeletal abnormalities	Regulation of blood levels of calcium and phosphate; immune function

Continued

TABLE 5.2.1 Vitamins—cont'd

Vitamin	Synonyms	Sources	Deficiency	Metabolic Role
Vitamin E	α-Tocopherol and other tocopherols; α-tocotrienol and other tocotrienols	Soy, corn, and olive oil; also nuts, seeds, wheat germ, cereals	Reduced sperm production	Antioxidant scavenger of reactive oxygen species
Vitamin K	Phylloquinone	Green leafy vegetables, vegetable oils, cereals	Subdural hemorrhaging	Required for carboxylation of glutamic acid in calcium binding blood clotting proteins

TABLE 5.2.2 Minerals

	Sources	Functions
Calcium	Milk, cheese, yogurt, sardines, canned fish with bones	Structural in bones and teeth, signal transduction
Phosphorus	Red meat, dairy foods, fish, poultry, bread, rice, oats	Structural in bones and teeth, DNA signal transduction, enzyme regulation
Sodium and chloride	Found in almost all foods	Fluid and electrolyte balance, nerve conduction
Magnesium	Spinach and Swiss chard; bran cereals and wheat germ; brown rice, dried beans, peas, and lentils such as black, navy, chickpeas, nuts and seeds such as almonds, cashews, pumpkin, sunflower and flax seeds	Nerve and muscle function; cofactor for many enzymes, including for DNA synthesis
Potassium	Bananas, papaya, sweet potato; dark leafy greens, avocado, prune juice, tomato juice, orange juice; milk, yogurt; dried beans such as navy, pinto and black beans, chickpeas, lentils, beef, pork, fish, nuts and seeds such as pistachio, almonds, pumpkin, flax and sunflower seeds	Fluid balance, nerve conduction, Blood pressure regulation
Sulfur	Sulfur is found in nearly all foods	Structural component in cartilage, antioxidant metabolism, important for protein function
Iron	Meat, fish, poultry, firm tofu, dried beans, peas, like soybeans, chickpeas, split pea, lentils, nuts and seeds, organ meats such as liver and heart; blackstrap molasses	Oxygen transport, signal transduction
Manganese	Tea, bread, nuts, cereals, green vegetables such as peas and runner beans	Enzyme cofactor
Copper	Nuts, shellfish	Blood cell formation, release of iron for hemoglobin function
Iodine	Fish and shellfish	Biosynthesis of thyroid hormones
Cobalt	Fish, nuts, green leafy vegetables, cereals	Component of vitamin B_{12}; amino acid metabolism and fatty acid metabolism

Continued

TABLE 5.2.2 Minerals—cont'd

	Sources	Functions
Zinc	Yogurt, milk, cheese; dried beans like kidney, navy, pinto and soybeans, lentils, pumpkin seeds and sunflower seeds, liver, meat, poultry, fish and seafood	Cofactor for enzymes, growth and development; immune function
Fluoride	Seafood, tea, gelatin	Fluoride helps prevent tooth decay
Boron	Green vegetables, fruits, nuts	Function in glucose metabolism and estrogen function
Selenium	Fish, shellfish, red meat, grains, eggs, chicken, liver, and garlic; vegetables grown in selenium-rich soil	Synthesis of antioxidant enzymes; immune function; cardiovascular support; bone and joint function; brain development

absorbed by the GI tract and are transported in blood in chylomicrons.

Vitamin A

Vitamin A is comprised of retinol, retinal, and retinoic acid which are classified as retinoids. The structure of retinol is representative in having a six-membered β-ionone ring with a double bond that is conjugated with the all trans double bonds of a polyunsaturated side chain derived from isoprene units and which bears a hydroxyl group (retinol; Figure 5.2.1), aldehyde group (retinal), a carboxyl group (retinoic acid), or a carboxylic acid derivative.

Vitamin A has numerous functions in humans. It is best known in the form of retinol for its function in forming rhodopsin, which is essential for vision. It is also important for growth and reproduction. In the form of retinoic acid, it has hormone-like activity. Good sources of vitamin A include liver and cod liver oil as well as cheeses, butter, carrot, and broccoli. Vitamin A can be acquired by conversion of the provitamin, β-carotene, which not surprisingly is a constituent of carrot as well as other vegetables such as tomatoes and red bell peppers, and contributes to the color of these vegetables. β-Carotene is a carotenoid, a class of compounds with many physiological functions in humans and other organisms. β-Carotene is cleaved by the enzyme β-carotene 15,15′-monooxygenase, which splits the β-Carotene in half and inserts oxygen to form two molecules of the aldehyde retinal for every β-Carotene converted.

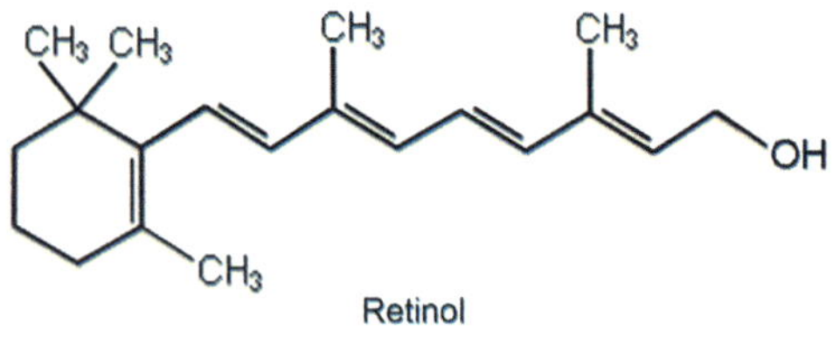

FIGURE 5.2.1 Structure of retinol.

Vitamin D

Vitamin D generally refers to vitamin D_2 and vitamin D_3 (Figure 5.2.2).

Vitamin D functions in a hormone-like manner in regulation of blood levels of calcium and phosphate and it is important for bone growth and physiology. Vitamin D is also involved in other physiological processes including immune function. The active form of vitamin D is 1,25-dihydroxycholecalciferol (1,25-$(OH)_2$ D_3). In the liver, vitamin D_3 is hydroxylated at carbon 25 to form 25-hydroxyvitamin D_3 (25-OH D_3),

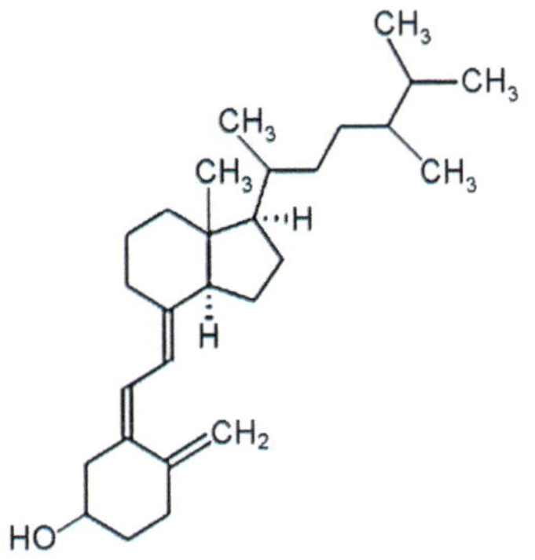

FIGURE 5.2.2 Structure of Vitamin D_3 (cholecalciferol).

which is secreted into the bloodstream and carried by vitamin D binding protein to the kidney where the second hydroxylation takes place at C1 in mitochondrial cells by the P-450 enzyme 25-hydroxyvitamin D-1α-hydroxylase. This second hydroxylation, which forms the physiologically active species, is highly regulated according to metabolic needs. 1,25-$(OH)_2$ D_3 action at the cellular level is mediated by its interaction with the vitamin D receptor (VDR) in cell nuclei, resulting in transcription of specific genes within effected cells. At least 14 genes have been identified to directly interact with the VDR.

Good sources of vitamin D_3 include liver, eggs, and fatty fish. Smaller amounts are found in butter and cheese. Shitake mushroom is a nonanimal source of vitamin D_2. Vitamin D_2 can be formed by the action of ultraviolet irradiation from sunlight acting on ergosterol. Ergosterol is a provitamin form of vitamin D_2 that is produced by fungi and yeasts. Vitamin D_3 can be made in the body from cholesterol after conversion to 5,7-cholestradienol, which is also known as 7-dehydrocholesterol. Conversion of 7-dehydrocholesterol to vitamin D3 occurs in skin. Upon irradiation with UVB ultraviolet rays from sunlight 7-dehydrocholesterol is converted to previtamin D3, which spontaneously isomerizes subsequently to form vitamin D3.

Vitamin E

There are 10 molecules under the category of vitamin E, which is comprised of tocopherols and tocotrienols. The most well known of the vitamin E, α-tocopherol exhibits the most potent physiological activity (Figure 5.2.3).

Alpha, beta, gamma, and delta tocopherols differ in the number and position of methyl groups on the six-membered ring. The six-membered aromatic ring is fused to a heterocyclic ring forming a chromane ring that bears an aliphatic carbon chain derived from isoprene units. The tocotrienols have a methylated chromane ring system and have three double bonds located within the aliphatic isoprene-derived side chain. α-Tocopherol deficiency has been shown to result in infertility in rats. The tocopherols and tocotrienols are potent antioxidants that effectively act as scavengers of reactive oxygen species and other free radicals, thereby preventing oxidative damage to cellular components.

Vitamin K

Vitamin K is essential for normal blood coagulation and some other calcium (Ca^{2+})-dependent functions. It is required for synthesis of prothrombin and some other coagulation factors that must bind Ca^{2+} ions to be functional. Vitamin K is involved in carboxylation of specific glutamic acid residues of these protein

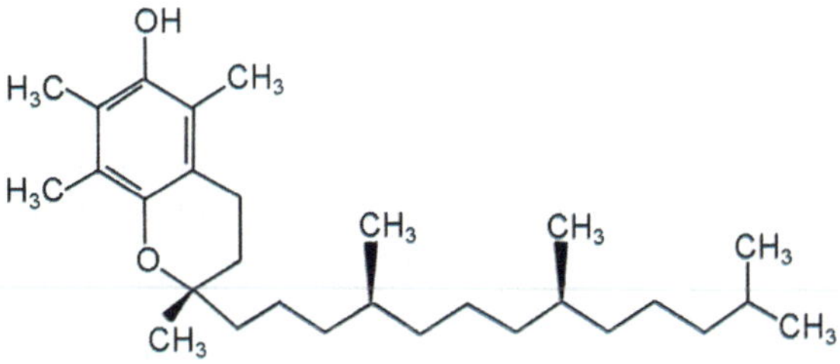

FIGURE 5.2.3 Structure of Vitamin E α-tocopherol

factors to form γ-glutamyl residues (gla residues), which have a negatively charged carboxylate group and improve binding of the proteins to Ca^{2+} in blood clotting. Animals fed warfarin, which has been used as a rat poison, synthesize an abnormal prothrombin that lacks gla and die from chemically induced hemorrhagic disease. Warfarin, also called Coumadin at appropriate dosage in humans, is an effective anticoagulant in prevention of thromboembolism.

WATER-SOLUBLE VITAMINS

Vitamin C

Vitamin C, L-ascorbic acid, and salts of its conjugate base, ascorbate, is required as a nutrient by six orders of primates (including humans), guinea pigs, and some bat and Passeriformes bird species. Loss of vitamin C biosynthetic activity in these animals is associated with mutation in the gene (GULO) that encodes the liver enzyme L-gulonolactone oxidase. It is believed that loss of GULO functionality occurred in human ancestors approximately 63 million years ago in part based on observations of Irwin Stone's research on ascorbic acid. Ascorbate is necessary for many metabolic reactions in plants and animals. Vitamin C has some antioxidant activity but its major role in human metabolism is as a cofactor for several enzymatic reactions. Biosynthesis of hydroxyproline, a component of collagen, requires vitamin C. Lack of sufficient vitamin C in the diet is associated with the deficiency disease scurvy. Some effects of scurvy are caused by defective secondary structure of collagen. Hydroxylation of proline to hydroxyproline occurs as a post translational modification and failure to accomplish this modification interferes with formation of collagen's triple helix structure. The recommended Dietary Allowance for vitamin C is 90 mg for men and 75 mg for women. Increased amounts are recommended for smokers and in pregnancy and lactation (Figure 5.2.4).

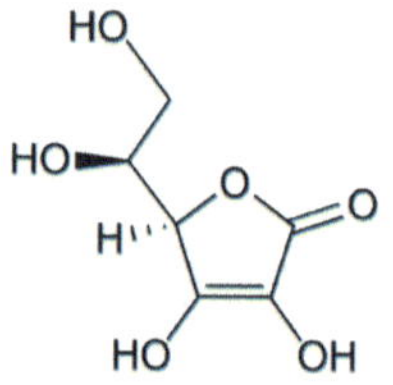

FIGURE 5.2.4 Structure of ascorbic acid.

Thiamin

Thiamin (also spelled thiamine), known as vitamin B_1, is found in meats (especially pork and organ meats), salmon, legumes, nuts, and grains. Thiamin, like vitamin C, is heat labile and loses activity during cooking. Deficiency of thiamin can cause beriberi and optic neuropathy and may play a role in the development of Korsakoff's syndrome. Thiamin action is associated with its phosphorylated forms, thiamin pyrophosphate (TPP) and thiamin triphosphate (TTP). TPP is a cofactor in decarboxylation reactions of alpha-keto acids including pyruvate decarboxylation by pyruvate dehydrogenase complex, which connects the Embden-Meyerhof pathway to oxidative phosphorylation by feeding acetyl-CoA into the Krebs cycle. TPP is also a cofactor in another reaction in energy metabolism, the alpha-ketoglutarate dehydrogenase reaction within the Krebs cycle. TPP is also required for leucine, isoleucine, and valine metabolism and by transketolase in the pentose phosphate pathway. The pentose phosphate pathway is important in glucose metabolism within the brain. Thiamin deficiency associated with excess alcohol consumption contributes to alcoholic brain injury. TTP is formed from TPP and it may have a role in activation of some chloride channels. TTP concentration in cells is regulated suggesting a role in mammalian tissues; however, its role is not currently understood.

Riboflavin

Riboflavin (B_2) is a component of the enzyme cofactors flavin adenine dinucleotide (FAD) and flavin mononucleotide (FMN). FMN and FAD

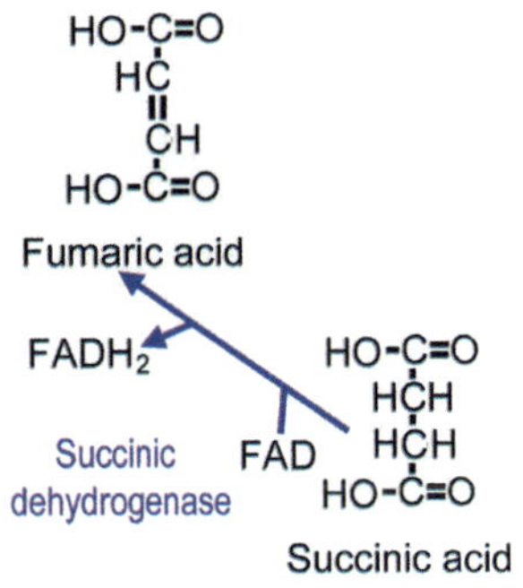

FIGURE 5.2.5 FAD activity.

function in a wide variety of reactions involving oxidation and reduction including electron transport, decarboxylation reactions, reduction of glutathione, fatty acid oxidation, and others. Many of the reactions involve the transfer of pairs of electrons such as in formation of double bonds in fatty acid desaturation reactions or the oxidation of succinate in the Krebs cycle (Figure 5.2.5).

Good sources of riboflavin include yeast, milk, cheese, green leafy vegetables, and nuts.

Panthothenic Acid

Pantothenic acid or vitamin B5 is a component of coenzyme A and phosphopantetheine, which is itself a component of coenzyme A. Phosphopantetheine is also a prosthetic group for acyl carrier protein, peptidyl carrier proteins, and aryl carrier proteins. Coenzyme A functions in forming thioesters with the carboxyl groups of fatty acids and other organic acids. Sources of pantothenic acid and derivatives include whole-grain cereals, legumes, eggs, meat, and yogurt (Figure 5.2.6).

Niacin

Niacin, also known as nicotinic acid or vitamin B_3, has the chemical name of pyridine-3-carboxylic acid. Nicotinamide is a component of nicotinamide adenine dinucleotide (NAD^+) and nicotinamide adenine dinucleotide phosphate ($NADP^+$). NAD^+ and $NADP^+$ are coenzymes that can accept two electrons to be reducibly reduced to NADH and NADPH. NAD and NADP participate in a wide assortment of redox reactions that are necessary for normal function in humans and other

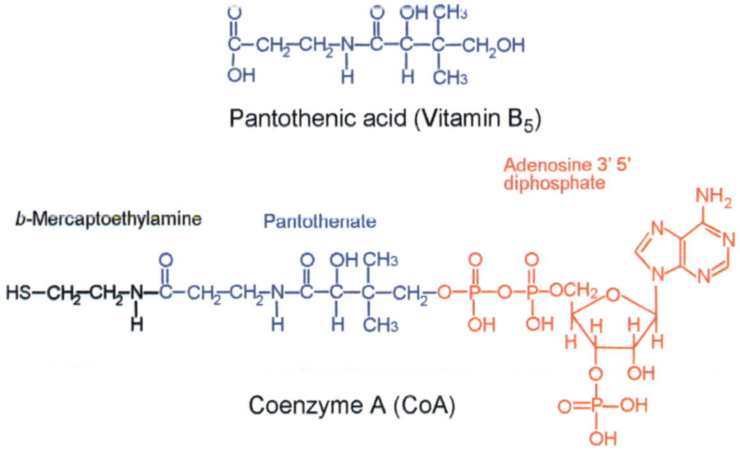

FIGURE 5.2.6 Structure of pantothenic acid and coenzyme A.

mammalian cells. In fact, NAD is an important cofactor for some of the most fundamental reactions in living systems. Niacin if consumed is converted to form the nicotinamide moiety of NAD after it is incorporated into the nicotinic acid analog (nicotinic acid mononucleotide) of NAD. Sources of niacin include fish and meats, liver, peanuts, mushrooms, sunflower seeds, and avocado. Niacin, though present in cereals, occurs as a glycoside, which is not hydrolyzed in the intestinal tract and is not absorbed. Nicotinamide from animal sources occurs as NAD and NADP. Free nicotinamide is released during digestion and absorbed. Niacin is not technically a vitamin because it can be synthesized in the human body from tryptophan but efficiency of conversion from tryptophan is low, about 1/60 of dietary tryptophan is converted. Chronic niacin deficiency results in pellagra, which has symptoms including diarrhea, dermatitis, dementia, delirium, inflammation of the mouth, and can lead to death if untreated.

Pyridoxine

Pyridoxine or vitamin B_6 is found in cereals, beans, vegetables, liver, meat, and eggs. The active form of the vitamin pyridoxal 5′-phosphate (PLP) is a cofactor for enzymes in a wide variety of metabolic reactions. Some of the reactions include transaminations, decarboxylations, and deaminations. Most but not all reactions requiring PLP involve its aldehyde functionality. In most reactions an aldimine forms between the aldehyde and an ε-amino group of a lysine residue in the enzyme's active site. The enzymic amino group is displaced by a substrate amino group to form an intermediate Schiff base. Subsequent reaction steps are particular to each type of reaction catalyzed.

Folic Acid

Folic acid or folate occurs in many foods including green leafy vegetables and other vegetables, fruits, nuts, meat, eggs, fish, grains, and shellfish. Deficiency is associated with physical and mental symptoms. Deficiency in pregnancy may lead to neural tube and brain defects in the developing fetus. Folates are involved in reactions where 1-carbon units are incorporated into biosynthetic products. These include reactions for DNA and RNA synthesis. Methylation of uracil to thymine is a particularly important reaction involving folate. Because cancer cells divide rapidly they synthesize DNA at higher rates than noncancer cells and are therefore more sensitive to disruption of folate-dependent reactions. Cancer chemotherapy often involves use of the folate analog methotrexate to selectively poison DNA synthesis in cancer cells (Figure 5.2.7).

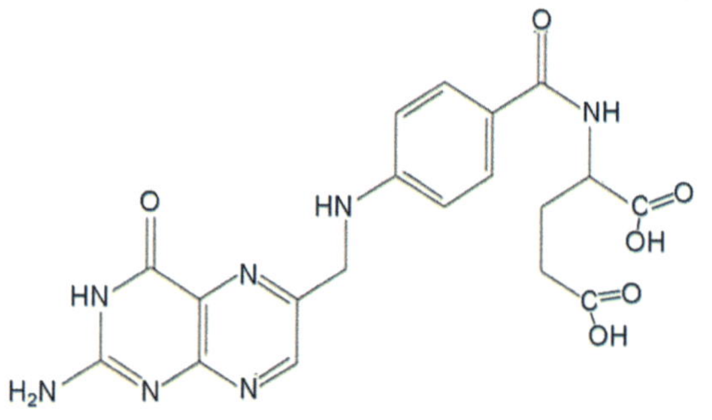

Folic acid

FIGURE 5.2.7 Folic acid structure.

Biotin

Biotin is also known as vitamin B7 or the anti-egg white injury factor. Biotin deficiency can be induced by feeding raw egg whites for prolonged periods. Deficiency caused by raw egg white consumption is due to the presence of the glycoprotein avidin, which binds biotin with high affinity. Biotin is comprised of an imidazole ring that has two nitrogen atoms. The imidazole ring is fused to a thiophene ring that in turn bears a 5-carbon valerate side chain. Biotin functions as a coenzyme in reactions catalyzed by several carboxylase enzymes. The carboxyl group of the valeryl side chain forms an amide linkage with an ε-amino group of a carboxylase

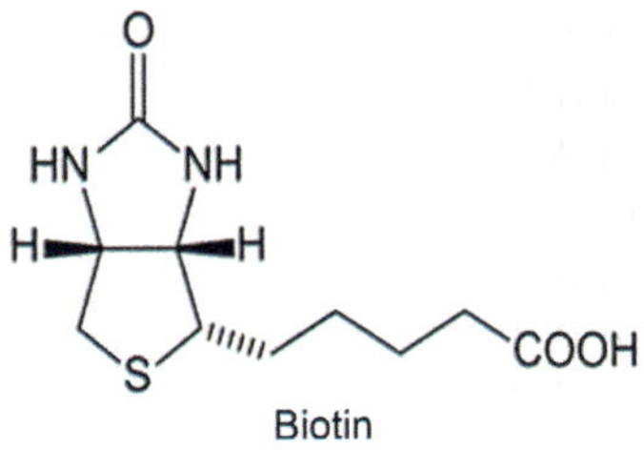

FIGURE 5.2.8 Biotin structure.

enzyme thereby covalently attaching the biotin molecule to the carboxylase enzyme. In humans there are four carboxylases that utilize biotin. These enzymes involve formation of an intermediate carboxybiotinyl enzyme in which the carbon from CO_2 is temporarily bound to one of the nitrogen atoms of the imidazole ring before transfer to its final destination. Biotin is found in many foods including egg yolk, liver, and yeast. Biotin is also produced in the human gut by bacteria. The Food and Nutrition Board recommends 30 micrograms (mcg) of biotin per day for adult intake (Figure 5.2.8).

Vitamin B_{12}

Vitamin B_{12} is required to form methylcobalamin and adenosylcobalamin, which are involved in two enzymatic reactions, the conversion of the nonproteinogenic amino acid homocysteine into methionine by methionine synthase and conversion of methylmalonyl-CoA to succinyl-CoA by methylmalonyl-CoA mutase. Vitamin B12 is found only in animal products and chlorella. Deficiency of B12, either due to insufficient intake or insufficient absorption, can cause pernicious anemia.

Adenosylcobalamin is required for methylmalonyl-CoA mutase activity. The enzyme catalyzes conversion of (*R*)-methylmalonyl-CoA to succinyl-CoA, a reaction in the pathway that handles entry of propionyl-CoA into the Krebs cycle. Propionyl-CoA is a thioester of the 3-carbon acid, propionic acid, and coenzyme A and it is formed from catabolism of isolucine, valine, threonine, methionine, thymine, cholesterol, and fatty acids with carbon chains with odd numbers of carbons. Recall from Chapter 1 that β-oxidation of fatty acids occurs by cyclic or repetitive removal of 2-carbon units from fatty acyl-CoA molecules. Successive removal of 2 carbons from a carbon chain with an odd number of carbons in these molecules eventually results in formation of propionyl-CoA. The 3-carbon propionyl moiety of propionyl-CoA is converted to a branched 4-carbon chain containing (*S*)-methylmalonyl-CoA by propionyl-CoA carboxylase. (*S*)-methylmalonyl-CoA is converted to its mirror image isomer (*R*)-methylmalonyl-CoA by methylmalonyl-CoA racemase. A defective enzyme is produced in individuals with an autosomal recessive disorder, methylmalonic acidemia (MMA), which is characterized by the inability to catalyze the conversion of methylmalonyl-CoA to succinyl-CoA. The disease is frequently fatal in the first year of life. The problem can be compounded because the symptoms of the disease are similar to ethylene glycol poisoning and at least one person had been arrested and convicted of murder when MMA had been misidentified as a case of ethylene glycol poisoning.

Minerals

Minerals are classified as major minerals or macrominerals if they are required in relatively large quantities as compared to trace minerals. Minerals are involved in diverse biological roles. The macrominerals include calcium, phosphorus, magnesium, sodium, potassium, chloride, and sulfur. Trace minerals include iron, manganese, copper, iodine, zinc, cobalt, fluoride, and selenium. Calcium and phosphate are major components of bone and teeth in the form of calcium hydroxyapatite. Calcium also functions in cell signaling and can also function as an enzyme cofactor in many reactions. Phosphate is also important in signal transduction and is involved in numerous signal transduction cascades. Many regulatory enzymes control fluxes through metabolic pathways when activated or deactivated by phosphorylation. Protein kinases are enzymes often involved in signal transduction and can attach a phosphate group to another protein, effectively switching it on or off. Iron functions with hemoglobin and myoglobin in binding oxygen and it also is a cofactor for some enzymes. The table below summarizes the major functions and sources of the macrominerals and trace minerals (Table 5.2.2).

CHAPTER KEYWORDS

Know these keywords and what each word refers to as well as its function. Read over the list. Are all these keywords familiar?

β-Carotene
1,25-Dihydroxycholecalciferol
Ascorbate
Biotin
Carotenoid
Cholecalciferol
Coenzyme A
FAD
Folate
Folic acid
Hypervitaminosis
Imidazole
Ascorbic acid
Methylcobalamin
Niacin
Nicotinic acid
Panthothenic acid
Provitamin
Pyridoxine
Retinal
Retinoic acid
Retinoids
Retinol
Riboflavin (B_2)
Scurvy
Thiamin
Thiamine
Tocopherols
Vitamin A
Vitamin B_1
Vitamin B_{12}
Vitamin B_3
Vitamin B_5
Vitamin B_6
Vitamin B_7
Vitamin C
Vitamin D
Vitamin D_2
Vitamin D_3
Vitamin E
Vitamin K
α-Tocopherol

INTERNET REFERENCES

Use these references to find out more.
en.wikipedia.org/wiki/Vitamin_A
en.wikipedia.org/wiki/B_Vitamins
en.wikipedia.org/wiki/Vitamin_C
en.wikipedia.org/wiki/Vitamin_D
en.wikipedia.org/wiki/Vitamin_E
en.wikipedia.org/wiki/Vitamin_K

CHAPTER QUIZ

Answer these questions. Do you know the answers, holy cow! Check yourself out in this quiz, and prove that you are a humanology wiz.

1. It is a true statement that vitamins
   - **a.** Provide energy for biochemical reactions
   - **b.** Are produced within the body within glandular organs
   - **c.** Frequently function as cofactors or coenzymes in enzymatic reactions
   - **d.** Are usually inorganic compounds
   - **e.** Cannot be synthesized in the laboratory
2. It is a true statement that vitamin C is
   - **a.** A fat-soluble vitamin found only in the skin of citrus fruit essential oils

- **b.** Associated with the deficiency disease pellagra
- **c.** A water-soluble vitamin that is stored in the liver
- **d.** Required for modification of proline in collagen
- **e.** A reactive oxygen species

3. It is a true statement that water-soluble vitamins

- **a.** Occur more abundantly in plants that absorb large quantities of water
- **b.** Do not build up in the body and must be ingested daily
- **c.** Are associated with hypervitaminosis
- **d.** Include only the B vitamins
- **e.** Are more stable than fat-soluble vitamins

4. It is a true statement that minerals are classified as

- **a.** Macrominerals if the molecular weight is very high
- **b.** Microminerals if the molecular weight is low
- **c.** Macrominerals if required in large quantities
- **d.** Trace minerals if required in minute quantities
- **e.** a and b are correct
- **f.** c and d are correct

CHAPTER

5.3

Diet and Body Weight

The diet should provide appropriate levels of macronutrients and sufficient levels of micronutrients to maintain good health of adults and to support growth and development in infants, children, and adolescents.

ENERGY BALANCE

Caloric content should be matched (calorie balance) to energy expenditure to avoid body weights that are overweight or underweight. The thermodynamic calorie is defined as the amount of energy required to raise the temperature of 1 g of water to 1°C. It should be kept in mind that the listing of dietary calories in nutrition tables and package labels for foods are expressed using the dietary calorie (Cal, with an uppercase C). One dietary calorie is equal to 1000 thermodynamic calories (1000 Cal) or 1 kilocalorie (1 kcal). Another unit of energy that is often encountered in the context of diet and exercise is the joule (J). One calorie (1 kcal) is equal to 4184 J. When using exercise equipment, people often find output expressed in watts (W) or kilowatts (kW). Watts are not directly convertible to Calories or joules because the watt is a unit of power (energy per unit time), whereas calories and joules are units of energy. Wattage × time (watt hours) yields energy that can be calculated in calories, joules, British Thermal Units, and so on. The number of calories expended using a device that lists wattage output by the user can be calculated using the equation below.

$$1\ \text{kW} \times \text{h} = 860\ \text{Cal}$$

$$1\ \text{W} \times \text{h} = 0.860\ \text{Cal}$$

For example, riding an exercise bicycle for 30 min at 100 W would use 100 W × ½ h = 50 Wh, and 50 Wh × 0.860 Cal/Wh = 43 Cal used. That is fewer calories than in half a slice of bread. This should illustrate that weight loss or prevention of excess weight gain cannot be achieved by exercise if caloric intake is excessive. This is also true for people who are trying to gain weight, such as bodybuilders and other athletes, who wish to build muscle rather than excess fat. Estimated daily caloric requirements are given in the table below assuming a reference man 5 ft 10 in. tall and weighing 154 pounds and assuming a reference woman 5 ft 4 in. tall and 126 pounds (Table 5.3.1).

Many practitioners in the health care and nutrition fields rely on body height-weight tables or various formulas to assess "ideal" body weight. It should be kept in mind that these are

http://dx.doi.org/10.1016/B978-0-12-803699-0.00038-4

TABLE 5.3.1 Estimated Daily Calorific Needs

	Estimated daily calorific needs					
Age	**Male sedentary**	**Male moderately active**	**Male active**	**Female sedentary**	**Female moderately active**	**Female active**
2	1000	1000	1000	1000	1000	1000
3	1200	1400	1400	1000	1200	1400
4	1200	1400	1600	1200	1400	1400
5	1200	1400	1600	1200	1400	1600
6	1400	1600	1800	1200	1400	1600
7	1400	1600	1800	1200	1600	1800
8	1400	1600	2000	1400	1600	1800
9	1600	1800	2000	1400	1600	1800
10	1600	1800	2200	1400	1800	2000
11	1800	2000	2200	1600	1800	2000
12	1800	2200	2400	1600	2000	2200
13	2000	2200	2600	1600	2000	2200
14	2000	2400	2800	1800	2000	2400
15	2200	2600	3000	1800	2000	2400
16	2400	2800	3200	1800	2000	2400
17	2400	2800	3200	1800	2000	2400
18	2400	2800	3200	1800	2000	2400
19-20	2600	2800	3000	2000	2200	2400
21-25	2400	2800	3000	2000	2200	2400
26-30	2400	2600	3000	1800	2000	2400
31-35	2400	2600	3000	1800	2000	2200
36-40	2400	2600	2800	1800	2000	2200
41-45	2200	2600	2800	1800	2000	2200
46-50	2200	2400	2800	1800	2000	2200
51-55	2200	2400	2800	1600	1800	2200
56-60	2200	2400	2600	1600	1800	2200
61-65	2000	2400	2600	1600	1800	2000
66-70	2000	2200	2600	1600	1800	2000
71-75	2000	2200	2600	1600	1800	2000
76+	2000	2200	2400	1600	1800	2000

Sedentary means a lifestyle that includes only the light physical activity associated with typical day-to-day life. Moderately active means a lifestyle that includes physical activity equivalent to walking about 1.5 to 3 miles per day at 3 to 4 miles per hour, in addition to the light physical activity associated with typical day-to-day life. Active means a lifestyle that includes physical activity equivalent to walking more than 3 miles per day at 3 to 4 miles per hour, in addition to the light physical activity associated with typical day-to-day life.
From USDA Dietary Guidelines for Americans 2010.

approximate methods and do not suit every individual perfectly. The most popular formulas were devised in large part to conform to the height-weight tables of the Metropolitan Life Company. The formula of BJ Devine was constructed in 1974 to approximate the Metropolitan Life tables and provides a simple calculation to estimate so-called Ideal Body Weight (IBW).

$$\text{Men } 50\,\text{kg} + 2.3\,\frac{\text{kg}}{\text{each inch over } 5\,\text{ft}}$$

$$\text{Women } 45.5\,\text{kg} + 2.3\,\frac{\text{kg}}{\text{each inch over } 5\,\text{ft}}$$

Similar formulas have been subsequently developed that allow slightly higher weights.

$$\text{Men } 52\,\text{kg} + 1.9\,\frac{\text{kg}}{\text{each inch over } 5\,\text{ft}}$$

$$\text{Women } 49\,\text{kg} + 1.7\,\frac{\text{kg}}{\text{each inch over } 5\,\text{ft}}$$

$$\text{Men } 56.2\,\text{kg} + 1.41\,\frac{\text{kg}}{\text{each inch over } 5\,\text{ft}}$$

$$\text{Women } 53.1\,\text{kg} + 1.36\,\frac{\text{kg}}{\text{each inch over } 5\,\text{ft}}$$

These formulas can be modified plus or minus 10% to accommodate persons with large or small frame sizes. Frame size can be determined by measuring the elbow breadth as widest part of the elbow in millimeters and height in centimeters. Elbow breadth can be determined using a caliper (Figure 5.3.1).

An index value is calculated according to the Frame Index formula.

$$\text{Frame Index} = \frac{\text{elbow breadth (mm)}}{\text{height (cm)}} \times 100$$

The frame classification is then looked up according to the Frame Size Classification Table under the appropriate age and gender (Table 5.3.2).

Body mass index (BMI) is also used to assess weight status. The major advantage of BMI is that it is easy to calculate and requires only measurement of height and weight. The formula for

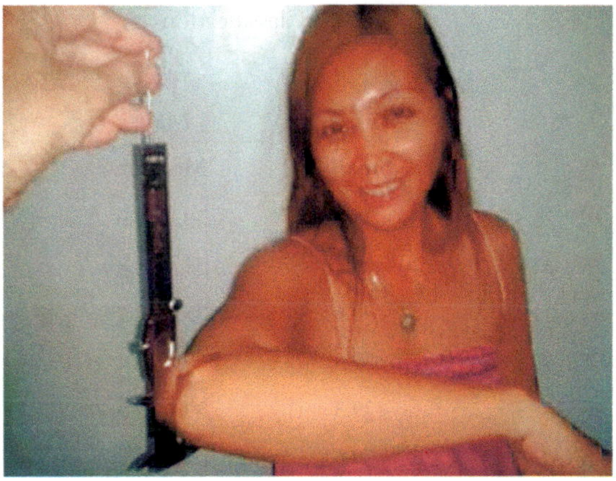

FIGURE 5.3.1 Elbow breadth measurements.

BMI is mass in kg divided by the square of height in meters. The ideal BMI on average is considered to be about 22, which is close to the middle of the normal range (Table 5.3.3).

$$\text{BMI} = \frac{\text{mass (kg)}}{\text{height}^2\,(\text{m}^2)}$$

DIET AND WEIGHT CONTROL

Body weight will increase, decrease, or remain constant depending upon energy balance. People may adjust their caloric intake to suit their needs; but all diets should contain adequate levels of vitamins, minerals, macronutrients, and essential components, such as essential fatty acids. A good rule of thumb when attempting to modify body weight is to do so gradually. Crash diets are not only dangerous but also usually doomed to failure. It is also essential to verify that prior to embarking on a weight loss or weight gain regimen to have a physical examination to assess whether calorie intake, total protein, and activity levels are safe. Because overweight and obesity are the most common issues relating to weight control, most of this

TABLE 5.3.2 Frame Size Classification

	Frame size classification					
	Male			Female		
Age	Small	Medium	Large	Small	Medium	Large
18.0-24.9	<38.4	38.4-41.6	>41.6	<35.2	35.2-38.6	>38.6
25.0-29.9	<38.6	38.6-41.8	>41.8	<35.7	35.7-38.7	>38.7
30.0-24.9	<38.6	38.6-42.1	>42.1	<35.7	35.7-39.0	>39.0
35.0-39.9	<39.3	39.1-42.4	>42.4	<36.2	36.2-39.8	>39.8
40.0-44.9	<39.3	39.3-42.5	>42.5	<36.7	36.7-40.2	>40.2
45.9-49.9	<39.6	39.6-43.0	>43.0	<37.2	37.2-40.7	>40.7
50.0-54.9	<39.9	39.9-43.3	>43.3	<37.2	37.2-41.6	>41.6
55.0-55.9	<40.2	40.2-43.8	>43.8	<37.8	37.8-41.9	>41.9
60.0-64.9	<40.2	40.2-43.6	>43.6	<38.2	38.2-41.8	>41.8
65.0-69.9	<40.2	40.2-43.6	>43.6	<38.2	38.2-41.8	>41.8

Values for Frame Size Classification taken from L. E. Matarese & M. M. Gottschlich. *Contemporary Nutrition Support: A Clinical Guide*, page 37 (Saunders 2003)

TABLE 5.3.3 Body Mass Index (BMI)

	BMI	Obesity class	Disease risk men 102 cm (40 in.) or less Women 88 cm (35 in.) or less	Disease risk men > 102 cm (40 in.) Women > 88 cm (35 in.)
Underweight	<18.5			
Normal	18.5-24.9			
Overweight	25.0-29.9		Increased	High
Obesity	30.0-34.9	I	High	Very high
	35.0-39.9	II	Very high	Very high
	40.0+	III	Extremely high	Extremely high

Disease risk for type 2 diabetes, hypertension, and CVD.
National Heart, Lung, and Blood Institute.

section will be focused on matters relevant to effective and safe weight reduction and maintenance of body weight once appropriate body weight has been achieved (Table 5.3.4).

The caloric requirement tables in this chapter and published in other sources that state caloric requirement as a function of physical activity, gender, and age can be misleading because they are somewhat oversimplified. About 60% of the body's energy expenditure is due to resting energy expenditure. The body's resting energy expenditure varies as the deviation of caloric balance increases. If it seems more difficult to lose weight after having lost a few pounds, that

TABLE 5.3.4 Energy Content of Macronutrients

Energy content of macronutrients	
Macronutrient	Energy content ($\frac{Cal}{g}$)
Protein	4
Fat	9
Carbohydrate	4

is because it is. One way to increase energy expenditure to offset the decline of resting energy expenditure caused by a weight reduction diet is through exercise. Exercise, in addition to providing health and mood benefits independent of weight, can help make any weight reduction diet more effective. In order to achieve safe gradual and tolerable/sustainable weight reduction, energy intake should not be less than about 60% of estimated daily caloric needs. For a 23-year-old moderately active woman, caloric energy expenditure would be expected to be about 2200 Cal. Reduction of caloric intake by 40% for weight reduction purposes would indicate a daily caloric intake of about 1320 Cal/da. This calculation assumes the person has a height of 5 ft 4 in. (163 cm) and has average frame size. To calculate the value for a woman who is 5 ft 10 in. (178 cm), we can increase the caloric intake proportionate to the difference in IBW of the person and the IBW of the reference person using any one of the IBW equations above. Assuming IBW = 45.5 kg + 2.3 × each inch over 5 ft, our reference person would have IBW of 54.7 kg and a 5 ft 10 in. woman would have an IBW of 68.5. Taking the ratio of the subject to the reference, we have a factor of $\frac{68.5}{54.7} = 1.25$. Multiplying this factor times 1320 Cal/da yields an allowable caloric intake of 1650 Cal per day for the tall moderately active woman of average frame dimensions. The 1650 Cal figure can be adjusted up or down by 10% to accommodate for a small frame or large frame. This 6 ft 2 in. (188 cm) author at the age of 21 reduced caloric intake to about 1500 Cal/day for about 8 months while walking at least one mile per day and went from 248 pounds (112 kg) to 180 pounds (82 kg) and is today, 42 years later 187 lbs (85 kg). Adoption of a permanent lifestyle change including regular exercise and sensible diet was the key element for the transformation.

For the general population, protein intake should be about $0.9\ \frac{g}{kg\ ideal\ body\ weight}$. Endurance athletes should consume about 1.3 g/kg and strength athletes should consume about 1.8 g/kg. Calorie reduction is most effectively achieved by reducing total carbohydrate and fat intake but maintaining adequate protein intake. This is particularly important for individuals who need to lose large amounts of weight because weight loss dieter can expect to lose some muscle mass. Muscle mass loss can be ameliorated in part by maintaining adequate protein intake. Fat intake should comprise from 15 to 35% of the total caloric intake. The remaining calories can be supplied by carbohydrates. For our 5 ft 10 in. woman who exercises moderately, the macronutrient quantities to effectively achieve weight loss can be calculated as follows.

First, the caloric intake of 1650 Cal/da is computed. Let us assume she prefers to adopt a diet where 30% of her caloric intake is fat calories. We can calculate protein in grams and fat in calories directly. $\text{Protein} = 0.9\ \frac{g\ protein}{kg\ IBW} \times 68.5\ kg\ IBW = 62\ g$ protein, which would contain 4 Cal/g totaling 248 Cal. Fat calories are 30% of 1650 = 495 Cal (at 9 Cal/g that would amount to 55 g of fat). The remaining calories should be carbohydrates. Carbohydrate calories can be calculated as total − fat calories − protein calories = 1650-495 − 247 = 908 carbohydrate calories. Using the energy content of carbohydrates of 4 Cal/g, 908 Cal is equivalent to 227 g of carbohydrates (Table 5.3.5).

The calculation is illustrated in the table below. Protein in grams and fat calories are entered first. Carbohydrate calories are then computed as

TABLE 5.3.5 Calorie Calculator

	Grams	Cal/g	Calories	Cal (%)
Protein	61.65	4	246.6	14.9
Fat (30% fat calories)	55	9	495	30.0
Carbohydrate	227.1	4	908.4	55.1

above. The rest of the values are calculated from the energy content (Cal/g) of each macronutrient. The right column is a computation of percentage of total energy contributed from each macronutrient. Our 5 ft 10 in. moderately active female subject should consume about 62 g of protein, 55 g of fat, and 227 g of carbohydrates.

The USDA recommends that proteins, fats, and carbohydrates should contribute the following percentages of energy content to the diet (Table 5.3.6).

The method for calculating the relative proportions of each macronutrient in grams illustrated here for our 5 ft 10 in. moderately active person produces macronutrient proportions with caloric values that fall within the recommended USDA guidelines while providing sufficient protein to maintain muscle mass and caloric content that will enable safe and efficient weight loss.

This method can also be used for a "low-carb" diet such as the Protein Power Diet. We will use a 6 ft 1 in. 35-year-old male with a large frame to calculate the macronutrient intake for this individual on a low-carb weight loss diet. The protein power diet is a popular low carbohydrate diet. Low carbohydrate diets have also been shown to be safe and effective in several studies. The weight loss portion of the diet consists of two phases: an initial phase with very low carbohydrate consumption (21-40 grams per day) and a second phase with a less extreme carbohydrate range (70-130 grams per day). An average moderately active 35-year-old man would require 2600 Cal estimated from the caloric needs table. For a large-frame man, this increases by 10% to 2860 Cal. The ideal body weight for the reference subject of 5 ft 10 in. is calculated to be $50\text{ kg}+2.3\times10=73\text{ kg}$. Our 6 ft 1 in. subject's IBW is 80 kg. Multiplying the ration 80/73 times 2860 Cal yields his caloric requirement of 3134 Cal. 60% of this for a weight loss diet is 1880 Cal. His protein requirement is $0.9\times80=72$ g. Let us calculate his macronutrient proportions using 35 g of carbohydrates per day. We begin by entering protein and carbohydrate mass in grams into our table (Table 5.3.7).

Fat calories are then computed from the total 1880 from which is subtracted the carbohydrate and protein calories to yield 1452 fat calories, a whopping 77% of energy consumption in fat calories (Table 5.3.8).

If desired, the protein might be increased to 110 g, having the effect of reducing fat calories to a still very high 69% (Table 5.3.9).

TABLE 5.3.6 USDA Recommendations

	Protein (Cal) [%]	Fat (Cal) [%]	Carbohydrate (Cal) [%]
Older children and adolescents	10-30	25-35	45-65
Adults	10-35	20-35	45-65

TABLE 5.3.7 Macronutrient Proportions

	Grams	Cal/g	Calories	Cal (%)
Protein	72	4	288	15.32
Fat		9		
Carbohydrate	35	4	140	7.45

TABLE 5.3.8 Calculating Fat Calories

	Grams	Cal/g	Calories	Cal (%)
Protein	72	4	288	15.32
Fat	161.3	9	1452	77.23
Carbohydrate	35	4	140	7.45

TABLE 5.3.9 Calculating Fat Calories

	Grams	Cal/g	Calories	Cal (%)
Protein	110	4	440	23.40
Fat	144.4	9	1300	69.15
Carbohydrate	35	4	140	7.45

At 72 or 100 g of protein per day, both low-carb diets have fat content well above and carbohydrate content below USDA guidelines. With relatively high fat intake and low carbohydrate intake, insulin secretion would decrease because of the reduced carb load. One effect in reducing insulin levels is to down regulate activity of the enzyme 3-hydroxy-3-methylglutaryl-CoA reductase (HMG-CoA reductase), which is a control point in cholesterol synthesis. By reducing HMG-CoA reductase activity, the effect of the low-carb diet is similar to that of statin drugs to reduce cholesterol synthesis in the liver from its building block, HMG-CoA. The net result is that even with high fat intake on a low-carb diet, beneficial effects on cholesterol and triglyceride levels are observed. Another effect of reducing insulin secretion is to increase lipolysis, which is the breakdown of fats in fat cells (adipocytes) to glycerol and fatty acids. The chemical equation for lipolysis is shown below.

$$\text{fat molecule} + 3H_2O \rightarrow \text{glycerol} + 3 \text{ fatty acid molecules}$$

$$\text{glycerol} - 3 - \text{phosphate} + 3 \text{ fatty acids} \rightarrow \text{fat molecule} + 3H_2O$$

In the fat cell, fat molecules are constantly being synthesized and broken down. Whether a fat cell becomes larger or smaller depends on which process predominates. In contrast to lipolysis, the synthesis of a fat molecule is not simply the reverse of lipolysis, but requires glycerol-3-phosphate to be combined with three fatty acid molecules.

$$\text{glycerol} - 3 - \text{phosphate} + 3 \text{ fatty acids} \rightarrow \text{fat molecule}$$

Fat cells cannot synthesize glycerol-3-phosphate from glycerol. Adipocytes synthesize glycerol-3-phosphate from breaking down glucose. Therefore, in the fat cell it takes sugar to make fat, and it requires insulin to permit entry of glucose into the fat cell. The effect of the low-carb diet reduces entry of glucose into the fat cell thereby reducing esterification of fats. The result is that lipolysis predominates and free fatty acids are oxidatively utilized. Some of the fatty acids liberated break down to Acetyl-CoA, which in a series of reactions produces HMG-CoA, which is then cleaved to form acetoacetate, a "ketone body," which also has a reduced form β-hydroxybutyrate. These ketones are preferentially utilized by many cells including cardiac cells as a source of energy. During starvation, the human brain can use ketones for up to 70% of its energy requirements. The key to the effectiveness of the low-carb diet is that it promotes lipolysis and fatty acid utilization while sparing muscle proteins from being broken down. Although a high carbohydrate low calorie diet will produce weight loss, the effect of high carbs is to increase glucose entry into fat cells making it easier for the fat cells to synthesize fats, leading to greater loss of muscle mass. Consequently, the weight loss dieter feels like he or she is working harder to achieve the desired result and may be disappointed by the loss of muscle mass.

CHAPTER KEYWORDS

Know these keywords and what each word refers to as well as its function. Read over the list. Are all these keywords familiar?

Body mass index
Body weight
Calorie
Energy balance
Elbow breadth
Frame index

Joule
Protein intake
Statin
Watt hours

INTERNET REFERENCES

Use these references to find out more.
en.wikipedia.org/wiki/Weight
www.webmd.com/diet/default.htm
en.wikipedia.org/wiki/Diet
www.webmd.com/diet/evaluate-latest-diets
en.wikipedia.org/wiki/Calorie

CHAPTER QUIZ

Answer these questions. Do you know the answers, holy cow! Check yourself out in this quiz, and prove that you are a humanology wiz.

1. Protein requirements:
   a. Are proportional to a person's height
   b. Depend upon a person's total body weight
   c. Depend upon a person's ideal body weight
   d. Depend upon a person's fat mass
   e. Depend upon a person's water-soluble vitamin status

2. Caloric requirements:
   a. Depend upon body weight only
   b. Depend upon ideal body weight and activity level
   c. Decrease if persons consume fewer calories than required for an extended period of time
   d. Decrease if persons exercise on an empty stomach
   e. b and c are correct

3. When a person reduces carbohydrate intake to 8% of caloric intake:
   a. Fats are converted into carbohydrates
   b. Insulin levels decrease and lipolysis increases
   c. Cholesterol synthesis increases
   d. Ketogenic amino acids are broken down to make glucose
   e. c and d are correct

4. Fat cells:
   a. Synthesize glucose from fatty acids
   b. Can break down fats but cannot esterify them
   c. Absorb glucose in response to glucagon increase
   d. Require glycerol-3-phosphate to make fat molecules
   e. Synthesize glycerol-3-phosphate from ATP and glycerol

5. In the human cholesterol is derived from:
   a. HMG-CoA building blocks
   b. Vegetable sources in the diet
   c. Animal sources in the diet
   d. Avocados and other tropical fruits
   e. a and c are correct

Section 6
Human Diseases and Treatments

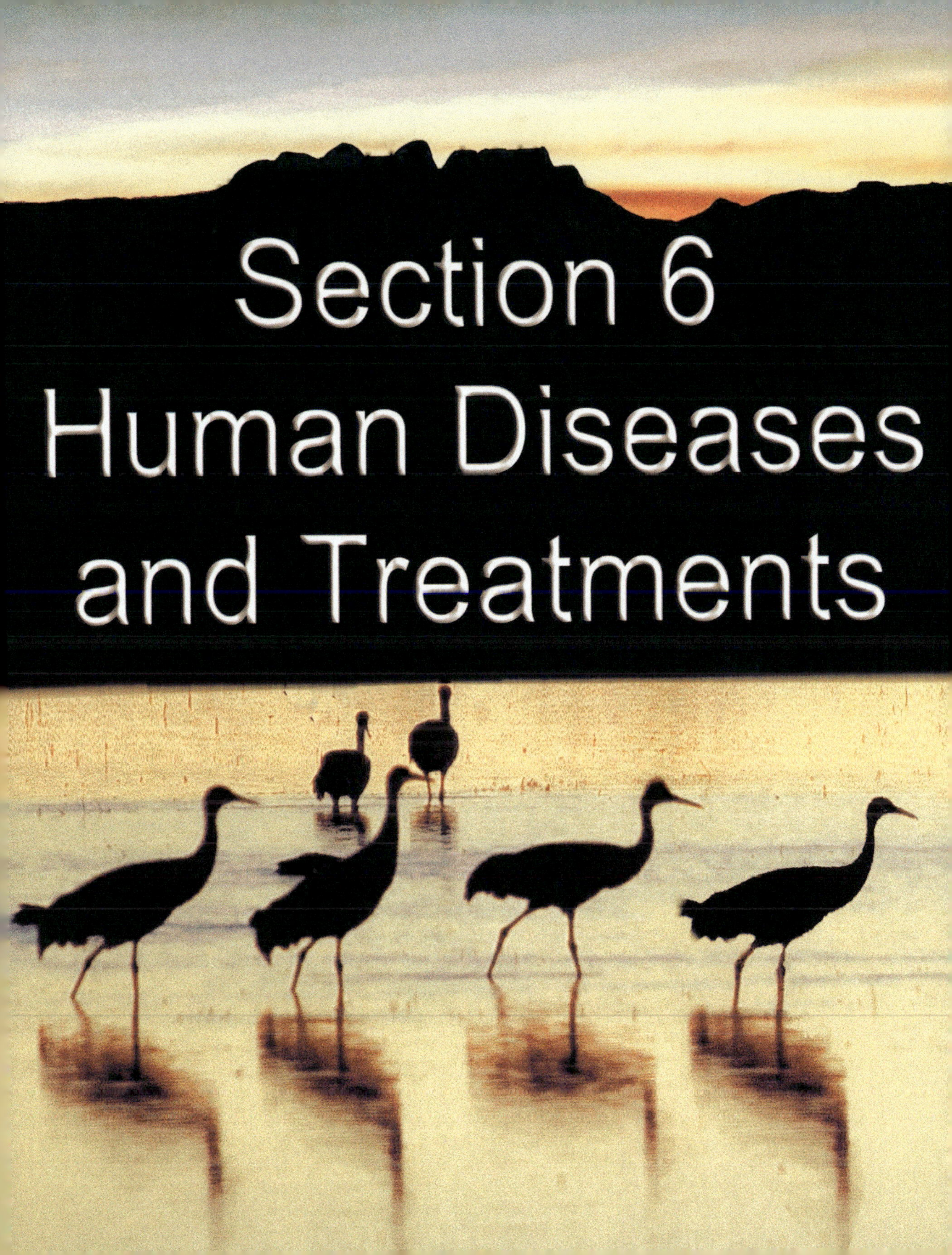

CHAPTER

6.1

Genetic Defects

Genetic disorders are quite common in humans, effecting approximately 1 in 70 new births in the United States. Both mothers and fathers can carry mutations or genetic faults in their genome. If, in contrast, a chromosome pair fails to separate properly during cell division, either at mitosis (cell division) or during meiosis (formation of a unique shuffled haploid or 23 chromosome set for reproduction), the egg or sperm may end up with a second copy of one of the chromosomes. If such a gamete results in fertilization and an embryo, the resulting embryo may also have an entire copy of the extra chromosome or a trisomy.

Genetic disorders may be a single gene disorder, resulting from a single mutated gene. Over 4000 known human diseases are caused by single gene defects. These disorders can be passed on to subsequent generations in multiple ways. For example, achondroplasia is typically considered a dominant disorder, but children with two genes for achondroplasia, one from the father and one from the mother, will have this severe skeletal disorder.

In an autosomal dominant genetic disorder, only one mutated copy of the gene, whether from the father or mother, is necessary for a person to be affected by the disorder. Examples of autosomal dominant genetic disorder include Huntingdon's Disease (Table 6.1.1).

In an autosomal recessive genetic disorder, two copies of the gene must be mutated, one from the father and one from the mother, for a person to be affected. Examples of this type of disorder are cystic fibrosis and sickle-cell anemia. Such disorders carry only a 25% chance with each pregnancy of an embryo carrying the disorder (Table 6.1.1).

X-linked dominant genetic disorders are caused by mutations in genes on the X chromosome. Some X-linked dominant conditions, such as Rett syndrome and Aicardi syndrome, are usually fatal in males either in utero or shortly after birth. A woman with an X-linked dominant disorder has a 50% chance of having an affected fetus with each pregnancy (Table 6.1.1).

X-linked recessive genetic conditions are also caused by mutations in genes on the X chromosome. Males are more frequently affected than females, and the chance of passing on the disorder differs between men and women. The sons of a man with an X-linked recessive disorder will not be affected while his daughters will carry one copy of the mutated gene. A woman who is a carrier of an X-linked recessive disorder has a 50% chance of having sons or daughters

Human Physiology, Biochemistry and Basic Medicine
http://dx.doi.org/10.1016/B978-0-12-803699-0.00006-2

TABLE 6.1.1 Prevalence of some examples of single gene disorders

Prevalence (number of live births)	
1. Autosomal dominant disorders	
Familial hypercholesterolemia	1 in 500
Polycystic kidney disease	1 in 1250
Neurofibromatosis type I	1 in 2500
Hereditary spherocytosis	1 in 5000
Marfan syndrome	1 in 4000
Huntington's disease	1 in 15,000
2. Autosomal recessive disorders	
Sickle cell anemia	1 in 625
Cystic fibrosis	1 in 2000
Tay-Sachs disease	1 in 3000
Phenylketonuria	1 in 12,000
Mucopolysaccharidoses	1 in 25,000
Lysosomal acid lipase deficiency	1 in 40,000
Glycogen storage diseases	1 in 50,000
Galactosemia	1 in 57,000
3. X-linked disorders	
Duchenne muscular dystrophy	1 in 7000
Hemophilia	1 in 10,000

who are affected. X-linked recessive conditions include the serious diseases hemophilia A and Duchenne muscular dystrophy (Table 6.1.1).

Y-linked genetic disorders are caused by mutations on the Y chromosome. These conditions may only be transmitted by males to offspring of the same sex.

Thalassemia is seemingly the most common genetic abnormality in the world. It effects as many as 1 in 30 pregnancies in China, India, and Pakistan, 1 in 100 pregnancies in Thailand, and 1 in 1000 pregnancies in Northern Europe and the United States. Thalassemia is an inherited blood disorder that can be categorized by abnormal formation of hemoglobin. The abnormal hemoglobin leads to irregular oxygen transport and destruction of red blood cells. Patients with thalassemia make less hemoglobin and have fewer circulating erythrocytes than normal, leading to mild or severe anemia.

Cystic fibrosis affects approximately 1 in 2000 pregnancies. Cystic Fibrosis is caused by the genetic lack of a specific protein. It leads to imbalance of the chloride ion in the body. Symptoms commonly include difficulty breathing, lung infections, and reproductive issues.

Sickle cell anemia affects 1 in 500 Black Americans or Africans, and 1 in 100,000 White Americans. It is a blood disorder characterized by an abnormality in hemoglobin. It leads to a propensity of erythrocyte cells that assume a rigid, sickle-like shape. Sickle-cell disease is associated with severe infections, attacks of severe pain, and an increased risk of death.

Down syndrome is a relatively common chromosome tripling abnormality in humans involving the tripling of chromosome 21 or trisomy 21. It affects approximately 1 in 700 pregnancies with the incidence rising with maternal age. The incidence may be as low as 1 in 5000 in a 20-year-old pregnancy, and as high as 1 in 100 in a 40-year-old pregnancy. Mosaics of Down syndrome occur; in these cases just some cells in the body have trisomy 21. In these cases, maternal cells have divided and we have just four root cells. In just one of these four root cells (divides into thousands of cells) is abnormally trisomy 21. Down syndrome is characteristically detected at birth by appearance of the face malformation. Problems include mental retardation, hearing and visual losses.

In Europe when Down syndrome is detected pre-birth 92% of parents opt for an abortion. In the United States, just 67% opt for termination of pregnancy. Opting for abortion may be difficult for Down syndrome parents. Children may be happy kids who are mentally retarded. They may not make it into a professional position, but may do simple work in a fast food chain, for

FIGURE 6.1.1 Two very happy kids with Down syndrome, trisomy 21 in Benin.

example. Figure 6.1.1 shows two happy kids with trisomy 21.

Trisomy 16 is an abnormality in which there are three copies of chromosome 16 rather than two. It is the most common trisomy in humans affecting 1 in 100 pregnancies. Trisomy 16 cannot lead to a live birth, and always results in miscarriage. It is possible for a child to be born alive with the mosaic form of trisomy 16. Like most chromosomal abnormalities, trisomy 16 usually causes miscarriage in the first trimester of pregnancy. Mosaic trisomy 16 births have mental retardation and heart defects.

Trisomy 18 (Edward's syndrome) and trisomy 13 (Patau's syndrome) derive from defective sperm-egg interactions and division of the fertilized gamete and lead to intellectual disability and microcephaly (small head and small brain). Other disabilities include heart and kidney defects. The majority of fetuses with trisomy 13 and 18 die before birth. The incidence of trisomy 13 is 1 in 10,000 pregnancies and the incidence of trisomy 18 is 1 in 6000 pregnancies.

There are numerous relatively common sex chromosome monosomies and trisomies. Monosomy X pregnancies or pregnancies with a single X chromosome (Turner's syndrome) occur at an incidence of 1 in 2000 pregnancies. This syndrome may manifest itself in a number of ways. Commonly, monosomy X women have short stature, low hairline, and low-set ears. Gonadal dysfunction or non-working ovaries is normal, which results in amenorrhea or absence of menstrual cycle and sterility. Finally, congenital heart disease occurs along with memory problems.

Triple X syndrome or trisomy X syndrome is a chromosomal disorder characterized by the presence of an extra X chromosome in each cell of a human female. Women with triple X syndrome have three X chromosomes instead of two. Triple X results during division of a parent's X and Y chromosomes and occurs about once in every 1000 births. Unlike other disorders such as Down syndrome, there is no distinguishable difference between those with triple X and the rest of the female population.

Trisomy X women are normally reproductively intact and able to bear children. Symptoms of trisomy X may include tall stature, microcephaly, and possibly language learning disabilities. Triple X females are at increased risk of poor academic performance and some may need special education. They benefit very much from a normal and steady home environment.

An XXY karyotype can exist; this is Klinefelter's syndrome, karyotype 47,XXY. Klinefelter's syndrome occurs in roughly 1 in 500 pregnancies. It is also due to an erroneous mitosis or meiosis in parent's X and Y chromosomes. A Klinefelter's baby will have small testicles and a small penis due to the presence of the Y chromosome and will be considered male at birth. Klinefelter's males and normal males may not be distinguished at birth. At puberty and during teens sterility may be recorded, and small testis, tall height, and gynecomastia (development of breasts) may be observed, leading to the identification of Klinefelter's syndrome. Klinefelter's syndrome men are commonly homosexuals.

Among trisomies, an XYY scenario also sometimes happens; it is XYY syndrome or 47, XYY males with an extra male Y chromosome. It occurs in roughly 1 in 1,000 male births. Some geneticists question whether the term "syndrome" is appropriate for 47 XYY because it is clinically normal, and most males do not know their karyotype.

The average IQ scores of XYY boys identified is not different compared to the general population. Approximately half of XYY boys identified have learning difficulties. Aggression is not seen more frequently in 47,XYY males, even though older studies suggested a correlation between 47,XYY males, aggression, and violence. The only medical symptom of 47,XYY syndrome is acne in adolescence, although behavior problems like impulsivity are suggested.

Other sex chromosome genetic disorders occur including SRY syndromes. SRY stands for Sex-related Region of Y chromosome. A group of genes coding for the development of the testis is located at the end of the short arm of the Y chromosome, which is considered the SRY zone. If during male meiosis this group of genes is translocated to the X chromosome, an XY fetus (no SRY) can be generated with no signal to generate testis, no testosterone, and no male characteristics (Figure 6.1.2). Phenotypically the child (46,XY-SRY) is recognized as a female having no penis. Alternatively, an XX fetus having the translocated SRY can have the signals to make testis, generate testosterone, and have a male or penis phenotype (46,XX+SRY). This baby will be identified as male (Figure 6.1.2).

Unfortunately both (46,XY-SRY) females and (46,XX+SRY) males are infertile, but have no other genetic or neurological problems. Studies with aborted fetuses and animals suggest that the occurrence of SRY syndromes (46,XY-SRY females and 46,XXSRY males) may be quite high, however, the incidence is a matter of conjecture. It has to simply be considered as a cause for infertility and a reason for homosexuality.

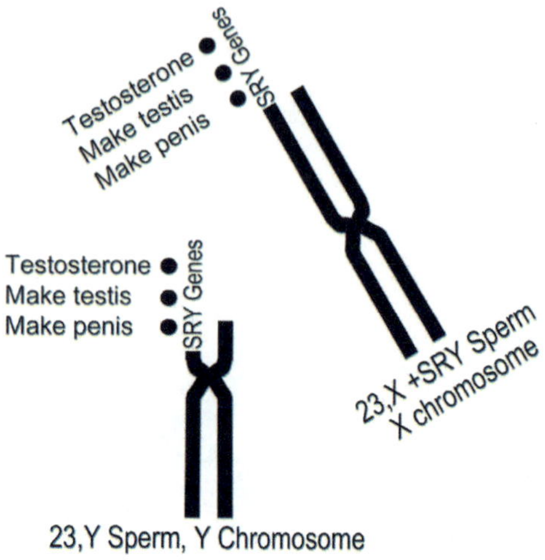

FIGURE 6.1.2 Meiosis Decision, to place the SRY genes on the X chromosome (makes 23,X+SRY sperm), or on the normal Y chromosome (makes 23,Y sperm).

We ask, is a 46,XY, 46 XY-SRY, 46,XX+SRY or 47,XXY, or 45,XO+SRY necessarily a male, and is a 46,XX, 46 XY-SRY, 46,XX+SRY or a 47, XXY, or 45,XO necessarily a female? In a day and age when homosexuality is in the spotlight these may be important questions.

CHAPTER KEYWORDS

Know these keywords and what each word refers to as well as its function. Read over the list. Are all these keywords familiar?

46,XX+SRY
46,XY-SRY
Achondroplasia
Aicardi syndrome
Cystic fibrosis
Down syndrome
Duchenne muscular dystrophy
Edward's syndrome
Gonadal dysfunction
Huntingdon's disease

Klinefelter's syndrome
Monosomy
Monosomy X
Patau's syndrome
Rett syndrome
Sickle-cell anemia
SRY
Tay-Sachs disease
Thalassemia
Trisomy 13
Trisomy 16
Trisomy 18
Trisomy 21
Trisomy X
Turner's syndrome
XXY
XYY

INTERNET REFERENCES

Use these references to find out more.
en.wikipedia.org/wiki/Down_syndrome
en.wikipedia.org/wiki/Genetic_disorder
en.wikipedia.org/wiki/Trisomy
www.nlm.nih.gov/medlineplus/ency/article/001660.htm
www.trisomy18.org/site/PageServer?pagename=whatisT18_whatis
en.wikipedia.org/wiki/Cystic_fibrosis
www.cff.org/AboutCF/
www.nhlbi.nih.gov/health/health-topics/topics/sca/
en.wikipedia.org/wiki/Sickle-cell_disease
ghr.nlm.nih.gov/chromosome/16
en.wikipedia.org/wiki/Trisomy_16
www.turnersyndrome.org/
www.trisomy16.org/
en.wikipedia.org/wiki/Turner_syndrome
ghr.nlm.nih.gov/condition/klinefelter-syndrome
www.genome.gov/19519068
en.wikipedia.org/wiki/XYY_syndrome
en.wikipedia.org/wiki/Testis_determining_factor
http://ghr.nlm.nih.gov/gene/SRY

CHAPTER QUIZ

Answer these questions. Do you know the answers, holy cow! Check yourself out in this quiz, and prove that you are a humanology wiz.

1. Humans normally have how many chromosomes?
 a. 47
 b. 46
 c. 48
 d. 44
 e. 23

2. Huntingdon's disease is an autosomal dominant condition. If your father has the Huntingdon's disease gene what is the chances that you as his son will get the disease?
 a. 12.5%
 b. 25%
 c. 50%
 d. 75%
 c. 100%

3. Edward's syndrome involves a tripling of which chromosome?
 a. 13
 b. 16
 c. 18
 d. 21
 e. X

4. Trisomies of the sex chromosome include?
 a. XXX, XXY, and XYY
 b. XXX and YYY
 c. XXY, XYY, and YYY
 d. XXX, XXY, XYY, and YYY
 e. XO, XXY, XXY, and YYY

5. A 46,XX+SRY fetus at birth will probably be identified by the delivering physician as being
 a. Male
 b. Female
 c. A hermaphrodite
 d. A homosexual
 e. An androgynous being

CHAPTER

6.2

Bacteria, Virus, Fungi, and Infectious Diseases

Some examples of bacterial, fungus, and viral infections are listed in Table 6.2.1. All can be aggressive life-threatening infections or nonhostile relatively mild short-term infections. In response to all infectious agents, the body generates antibodies or generates resistance to infection agent. In all cases, bacteria, fungi, and viruses infect one cell or one tissue and then start to spread around the body infecting multiple other cells or tissues, in some cases killing humans. One common symptom of most bacterial, fungal, and virus infections is fever.

Fever or temperature is regulated in the hypothalamus of the brain. A trigger of the fever, called a pyrogen, causes a release of the large folded lipid prostaglandin E2. Prostaglandin E2 acts on the hypothalamus causing heat-creating effects to match a new temperature level. The hypothalamus works like a thermostat. When the set point is raised, the body increases its temperature through both active generation of heat and retaining heat. Vasoconstriction reduces heat loss through the skin and causes the person to feel cold. If these measures do not make the blood temperature in the brain match the new setting, then shivering begins in order to use muscle movements to produce heat.

A pyrogen is a substance that induces fever. These can be either internal or external to the body. The bacterial sugary fat substance lipopolysaccharide, present in the cell wall of bacteria, is an example of a pyrogen. In extreme examples bacterial pyrogens can themselves cause rapid and dangerous fevers. All pyrogens, bacteria, fungi, and viruses produce cytokines, molecules that are a part of the immune system. These can cause the increase in the thermoregulation set point in the hypothalamus.

Symptoms of infection depend very much on the type of disease. General symptoms of infection include fatigue, loss of appetite, weight loss, fevers, night sweats, chills, aches, and pains. The classic symptoms of a bacterial infection are local redness, fever, swelling, and pain at the site of infection. In general, viral infections are systemic, involving many different parts of the body at the same time; i.e., a runny nose, cough, and body aches.

Antibacterial, antimycotic, and antiviral drugs have been developed that generally function by

Human Physiology, Biochemistry and Basic Medicine
http://dx.doi.org/10.1016/B978-0-12-803699-0.00040-2

TABLE 6.2.1 Examples of Some Bacterial, Fungal, and Virus Infections

Pneumonia	Pneumococci	Bacteria
	Streptococci	Bacteria
	Rhinoviruses	Viruses
	Coronaviruses	Viruses
Meningitis	Meningococci	Bacteria
	Streptococci	Bacteria
	Enterovirus	Viruses
	Herpes simplex type 2	Viruses
	Cryptococcal	Fungi
Mycosis	Tinea versicolor	Fungi
Gonorrhea	Neisseria gonorrhoeae	Bacteria
Gastic ulcer	Helicobacter pylori	Bacteria
Syphilis	Treponema pallidum	Bacteria
Lyme disease	Borrelia burgdorferi	Bacteria
Encephalitis	Herpes simplex	Viruses
	Flavivirus	Viruses
Common Cold	Rhinovirus	Viruses
Influenza	Influenza virus	Viruses
HIV/AIDS	Lentivirus	Viruses
Hepatitis	Hepatitis B virus	Viruses
Ebola hemorrhagic fever	Ebolavirus	Viruses

blocking bacteria, fungi, and viruses from replicating and surviving. Penicillin as an antibacterial, for instance, functions by inhibiting synthesis of bacterial cell walls. Examples of penicillin drugs include amoxicillin and ampicillin. Bacteria constantly try to rebuild their peptidoglycan cell wall. Penicillin blocks the formation of peptidoglycan cross-links in the cell wall. The four-membered β-lactam ring of penicillin binds to the enzyme transpeptidase in bacteria. As a consequence, transpeptidase cannot catalyze formation of these cross-links. The imbalance between cell wall production and degradation causes the bacteria cells to rapidly die.

Streptomycin or streptomycin sulfate is an antibacterial; it is a ribosome protein synthesis inhibitor. It binds to the small ribosomal RNA (rRNA), 30S component of the bacterial ribosome causing codon misreading, eventual inhibition of protein synthesis and bacterial replication, and ultimately death of bacterial cells. Humans have structurally very different ribosomes from bacteria, thereby allowing the selectivity of this antibiotic for bacteria. Tetracycline is an antibacterial that also functions by binding the small, 30S, bacterial ribosome, and blocking codon reading and bacterial replication.

A third type of antibacterial is fluoroquinolones. These include drugs such as Cipro and Levaquin. These are synthetic, bactericidal antibacterial agents with broad-spectrum activity. They inhibit the enzyme topoisomerase II, a DNA gyrase that is necessary for the replication of the microorganism. Topoisomerase II enzyme produces a negative supercoil on DNA, permitting transcription or replication and thereby inhibiting enzyme DNA replication and transcription. Blocking transcription blocks any possibility of replication.

Unlike most antibiotics, antiviral drugs do not destroy their target microbe; in its place they inhibit their progress and development of viruses. Antivirals should be distinguished from viricides, which are usually used a disinfectants rather than medications, but deactivate or destroy virus particles, either inside or outside the body. One modern antiviral strategy is to design drugs to interfere with a virus's ability to infiltrate a target cell. The virus must go through a sequence of steps to do this, starting with binding to a specific "receptor" molecule on the surface of the host cell and ending with the virus "uncoating" inside the infiltrated cell and releasing its contents.

A very early stage of viral infection is viral entry, when the virus attaches to and enters the host cell. A number of "entry-inhibiting" or "entry-blocking" drugs are being developed to fight virus infection. Inhibitors of uncoating have also been investigated. Amantadine and rimantadine have been introduced to combat influenza. These agents act on this penetration and uncoating process. Pleconaril works against rhinoviruses, which cause the common cold, by blocking a pocket on the surface of the virus that controls the uncoating process. This pocket is similar in most strains of rhinoviruses and enteroviruses, which can cause diarrhea, meningitis, conjunctivitis, and encephalitis.

The problem today with antibiotics or antibacterials, antifungicides, and antiviruses is resistance. Antimicrobial resistance is resistance of a microorganism to an antimicrobial drug that was originally effective for treatment of infections caused by it.

Resistant microorganisms (bacteria, fungi, viruses, and parasites) are able to withstand attack by antimicrobial drugs so that standard treatments become ineffective and infections persist, increasing the risk of spread to others. The evolution of resistant strains is a natural phenomenon that occurs when microorganisms replicate themselves erroneously or when resistant traits are exchanged between them.

When resistance occurs it is necessary to form modified drugs to avoid resistance. For instance, ampicillin and amoxicillin are variants of penicillin to work around resistance of common infections.

CHAPTER KEYWORDS

Know these keywords and what each word refers to as well as its function. Read over the list. Are all these keywords familiar?

Cipro
Cytokine
Ebola hemorrhagic fever
Encephalitis
Fluoroquinolones
Gonorrhea
Hepatitis
HIV/AIDS
Hypothalamus
Meningitis
Mycosis
Pneumonia
Prostaglandin E2
Pyrogen
ß-lactam
Streptomycin
Syphilis
Tetracycline
Topoisomerase
Transpeptidase
Uncoating

INTERNET REFERENCES

Use these references to find out more.
en.wikipedia.org/wiki/Antibiotics
en.wikipedia.org/wiki/Antiviral_drug
en.wikipedia.org/wiki/Infection
en.wikipedia.org/wiki/Fever
en.wikipedia.org/wiki/Penicillin
en.wikipedia.org/wiki/Quinolone
en.wikipedia.org/wiki/Streptomycin

CHAPTER QUIZ

Answer these questions. Do you know the answers, holy cow! Check yourself out in this quiz, and prove that you are a humanology wiz.

1. Penicillin kills bacteria and prevent them from replicating by:
   a. Binding their 30S ribosome and blocking protein production
   b. By binding their DNA topisomerase and blocking it

c. Binding the key proteoglycan blocking replication
d. Binding the cell isomerase receptor and blocking cell replication
e. Binding the membrane transpeptidase and preventing cell wall formation

2. Syphilis is caused by Treponema Pallidum, which is a:
a. Virus
b. Fungus
c. Bacterium
d. Protozoan
e. Prion

3. Fever is generally caused by:
a. The hypothalamus
b. The thyroid
c. The kidney
d. Blood macrophages
e. Blood T-cells

4. Streptomycin kills bacteria and prevent them from replicating by:
a. Binding their 30S ribosome and blocking protein production
b. Binding their DNA topisomerase and blocking it
c. Binding the key proteoglycan blocking replication
d. Binding the cell isomerase receptor and blocking cell replication
e. Binding the membrane transpeptidase and preventing cell wall formation

CHAPTER

6.3

Human Cancers and Carcinogenesis

By definition cancer is the uncontrolled growth of abnormal cells in the body, with the potential to spread to other parts of the body. Cancerous cells are also called malignant, spreading, or deadly cells. Carcinogenesis is the process whereby cancer starts. Some cancers are deadly, like pancreatic adenocarcinoma, with little chance of survival. Other cancers like skin basal cell carcinoma have only a tiny chance of death. Over one hundred different cancers affect humans.

Cancer is caused by exposing human cells to undesirable chemicals, drugs, foods, and agents including ultraviolet light and radiation. Cancer is the result of a chemically damaged cell's fight for survival. Tobacco use is the cause of 22% of cancer deaths. It is estimated that 10% of cancer deaths are due to obesity or poor diet. In 2012, 14.1 million new cases of cancer were reported globally, and 8.2 million people died from cancer.

The vast majority of cancers or 90-95% of cases are due to environmental factors. The remaining 5-10% are due to inherited cancer genetics. Environmental factors that contribute to cancer deaths include tobacco, diet and obesity, infections, stress, lack of physical activity, and environmental pollutants.

Carcinogenesis is the creation of cancer. It is a pathway by which normal cells are transformed into cancer cells, characterized by a progression of changes at the cellular, genetic, and epigenetic (DNA-associated regulatory histones) level that reprograms a cell to undergo uncontrolled cell division, or to become a cancerous mass. Cell division is a physiological process occurring in almost all tissues. Mutations and epimutations (modifications of DNA-associate histones) in DNA lead to cancer. Only certain mutations and epimutations lead to cancer, but the majority of mutations and epimutations have no such effect.

Research shows that carcinogenesis is caused by mutations and epimutations of the genetic material of normal cells upsetting the normal balance between proliferation and apoptosis or cell death. More than one mutation is needed for carcinogenesis to occur. In fact, a series of mutations to certain classes of genes is required before a normal cell will become a cancer cell. In particular, mutations in those genes that play vital roles in cell division, apoptosis, and DNA repair will cause a cell to lose all control. Vital genes from a cancer perspective include proto-oncogenes, genes that when activated, or mutated or perverted, can become cancer-promoting genes, and tumor suppressor genes, genes that normally inhibit cell division until

Human Physiology, Biochemistry and Basic Medicine
http://dx.doi.org/10.1016/B978-0-12-803699-0.00041-4

mutated and permit wild cell growth when corrupted. Typically, changes in many genes including proto-oncogenes and tumor suppressor genes are required to transform a normal cell into a cancer cell.

Large-scale mutations that occur in cancer cases involve the deletion or gain of a portion of a chromosome. Genomic amplification occurs when a cell gains many copies, often 20 copies or more, of a small chromosomal region, usually containing one or more oncogenes and adjacent genetic material. Small-scale mutations include point mutations, deletions, and insertions, which may occur in the promoter of a gene and affect its expression, or may occur in the gene's coding sequence and alter the function or stability of its protein product. Epimutations include methylations or demethylation of the promoter regions of genes. Epimutations can also occur by acetylation or methylation or other alterations to histones, creating histones that repress or activate gene expression.

In general, today's research suggests two models for carcinogenesis as illustrated in Figures 6.3.1 and 6.3.2: A model based around proto-oncogenes causing carcinogenesis (Figure 6.3.1), and a larger model examining mutations of tumor suppressor gene, DNA repair genes, and activation of proto-oncogenes (Figure 6.3.2).

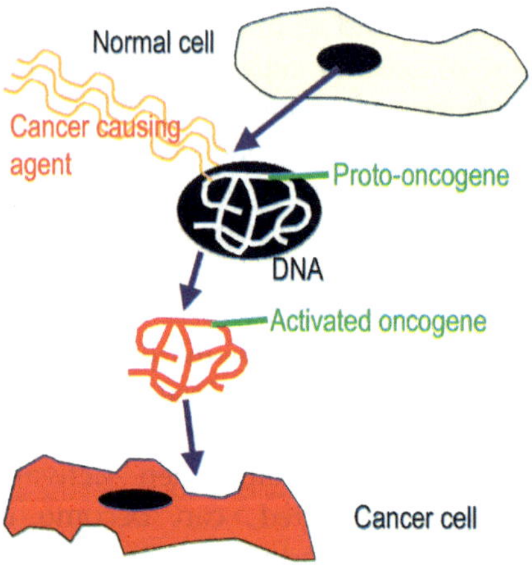

FIGURE 6.3.1 Model of carcinogenesis.

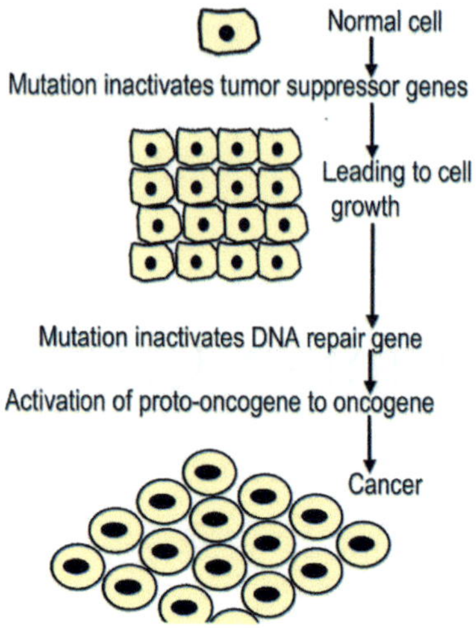

FIGURE 6.3.2 Model of carcinogenesis.

Most cancer research shows involvement of polypeptide growth factors in cancer cell growth, blockage of apoptosis, and invasion. These are transforming growth factor (TGF)-α, TGF-ß, platelet-derived growth factor (PDGF), and nerve growth factor (NGF). A new cancer cell model that uses TGFß, the most promising promoter of cancer cells, is the human chorionic gonadotropin free ß-subunit (hCG free ß-subunit) model. In this model, cancer cells produce hCG free ß-subunit, i.e., regular free ß-subunit and hyperglycosylated hCG free ß-subunit.

Humans have as part of their genome hyperglycosylated hCG as a molecule that functions in pregnancy to drive placental cytotrophoblast cell growth, blocks cytotrophoblast cell apoptosis to promote growth further, and promotes cell production of two cell-cell invasive enzymes, collagenase, and metalloproteinases to permit cell-cell invasion or the process of imbedding cytotrophoblast cells deep in the uterus (see Chapter 3.10). Hyperglycosylated hCG is a highly evolved ultra-super-duper promoter of this process (Chapter 4.1). It functions by

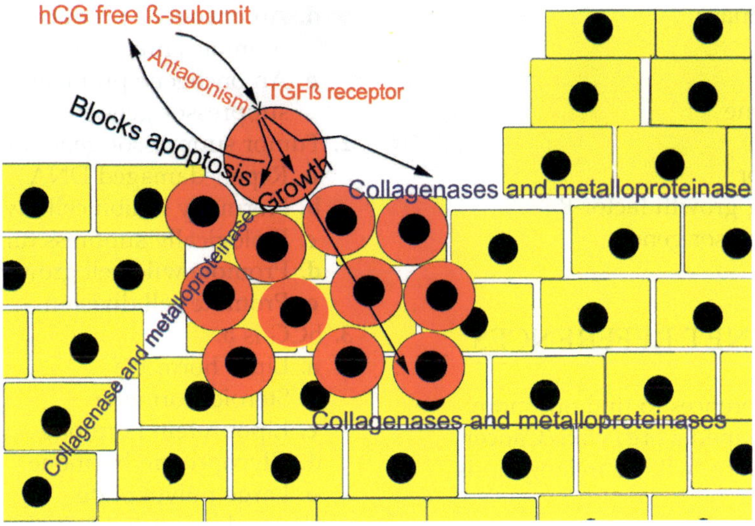

FIGURE 6.3.3 Action of hCG free ß-subunit on cancer cells.

antagonizing a TGFß receptor, leading to blockage of apoptosis and enzymatic cell-cell invasion (Figure 6.3.3).

Human cancers seem to steel this ultra-super-duper growth and invasion promoter, steal this entire implantation pathway, and use it to drive cancers hard. It is as though cancer cells know that the body has the ultra-super-duper process in its genome and steal it accordingly. Carcinogenesis seemingly involves activation of an hCG ß-subunit gene on chromosome 19, seemingly by mutation or epimutations of the surrounding histones. The complete folding of hCG in cytotrophoblast cells requires two separate disulfide-forming isomerases unique to the placenta. Cancer cells lack this enzyme. The result is that without this isomerase, ß-subunit s not properly formed, making a free ß-subunit and not the hCG α- ß-subunit dimer. Apparently, the free ß-subunit has similar biological activity at the TGFß receptor as hyperglycosylated hCG, promoting growth, blocking apoptosis, and promoting cell-cell invasion (Figure 6.3.3).

Seven independent research groups have now shown that this free ß-subunit functions in cancer cells through the pregnancy implantation pathway, promoting cancer cell growth, blocking apoptosis promoting growth further, and promoting cell-cell invasion. To date, research shows hCG free ß-subunit as the driver in bladder cancers, endometrial cancers, cervical cancers, breast cancer, ovarian cancers, Kaposi sarcoma, testicular cancers, and choriocarcinoma. While all cancers have not been investigated, other cancer growth and invasion pathways may exist.

CHAPTER KEYWORDS

Know these keywords and what each word refers to as well as its function. Read over the list. Are all these keywords familiar?

Apoptosis
Cancerous
Carcinogenesis
Collagenase
DNA repair gene
Epimutation
hCG
hCG free ß-subunit
Histone
Human chorionic gonadotropin

Metalloproteinase
Mutation
Oncogene
Proto-oncogene
TGFß
TGFß receptor
Transforming growth factor-ß
Tumor suppressor gene

INTERNET REFERENCES

Use these references to find out more.
http://en.wikipedia.org/wiki/Cancer
en.wikipedia.org/wiki/Carcinogenesis

CHAPTER QUIZ

Answer these questions. Do you know the answers, holy cow! Check yourself out in this quiz, and prove that you are a humanology wiz.

1. A proto-oncogene is:
   - **a.** A gene that promotes cancer synthesis
   - **b.** A protein involved in cancer mutations
   - **c.** A gene that when activated leads to cancer cell formation
   - **d.** An oncogene produced by cancers that promotes mutation
   - **e.** An oncogene promoted by tumor suppressor gene
2. Tumor suppressor genes are genes that:
   - **a.** Repair damaged DNA
   - **b.** Normally inhibit cell division
   - **c.** Biologically suppress tumors
   - **d.** Promote wild cell growth
   - **e.** Promote cell division
3. hCG is a:
   - **a.** Lipid hormone
   - **b.** Steroid hormone
   - **c.** Lipoprotein hormone
   - **d.** Glycoprotein hormone
   - **e.** Large polypeptide
4. Hyperglycosylated hCG functions by acting:
   - **a.** As an antagonist on a TGFß receptor
   - **b.** As an agonist on a TGFß receptor
   - **c.** On an NGF receptor
   - **d.** On a tumor suppressor protein receptor
   - **e.** On and hCG/LH hormone receptor
5. Epimutations are mutations:
   - **a.** In DNA sequence involving DNA epimers
   - **b.** That effect histones associated with DNA
   - **c.** In DNA complementary epi-sequence
   - **d.** In DNA exons
   - **e.** In DNA introns

CHAPTER

6.4

Cardiovascular Disease

Cardiovascular disease refers to any disease that affects the cardiovascular system; this includes coronary heart disease, myocardial infarction or heart attack, stroke or cerebrovascular accident (CVA), vascular diseases and/or peripheral arterial diseases. Cardiovascular diseases are the leading cause of deaths worldwide. Although cardiovascular diseases primarily affect older adults, atherosclerosis, beginning in early life, affects children and youths. Efforts should be directed on preventing atherosclerosis by modifying risk, for example by healthy eating, exercise, and avoiding tobacco use.

Research suggests a number of risk factors for cardiovascular diseases, including age, gender, high blood pressure, hyperlipidemia, diabetes, tobacco smoking, processed meat consumption, excessive alcohol consumption, sugar consumption, family history of cardiovascular disease, obesity, and lack of physical activity. Age is the most important risk factor in developing cardiovascular diseases, with approximately a tripling of risk with each decade of life. It is estimated that 82% of people who die of coronary heart disease are 65 and older. The risk of stroke or CVA doubles every decade a person ages after age 55. Figure 6.4.1 shows recent incidence and death statistics published by the American Heart Association and the American Stroke Association.

Research indicates that serum cholesterol levels may explain why age increases the risk of cardiovascular heart diseases. In most populations, the serum total cholesterol level increase with age. In men, levels increase to around age 45-50 years, in women, levels increase continuously until age 60-65 years. Aging is also associated with changes in the mechanical properties of the vascular wall, leading to loss of arterial elasticity. This loss of elasticity may subsequently lead to coronary artery disease.

Men are at greater risk of coronary heart disease than women prior to menopause. Once past menopause, men and women have similar risk factors. Among younger people, coronary heart disease is 2-5 times more common in men than in women. Among men and women, there are major differences in body weight, height, and body fat distribution. In the elderly, large artery pulsatility and stiffness are more pronounced in women. This may be caused by the smaller body size and arterial dimensions independent of menopause.

Cardiovascular diseases include coronary artery disease, also known as coronary heart disease, ischemic heart disease, and myocardial infarction. All involve blockage of a coronary

Human Physiology, Biochemistry and Basic Medicine
http://dx.doi.org/10.1016/B978-0-12-803699-0.00042-6

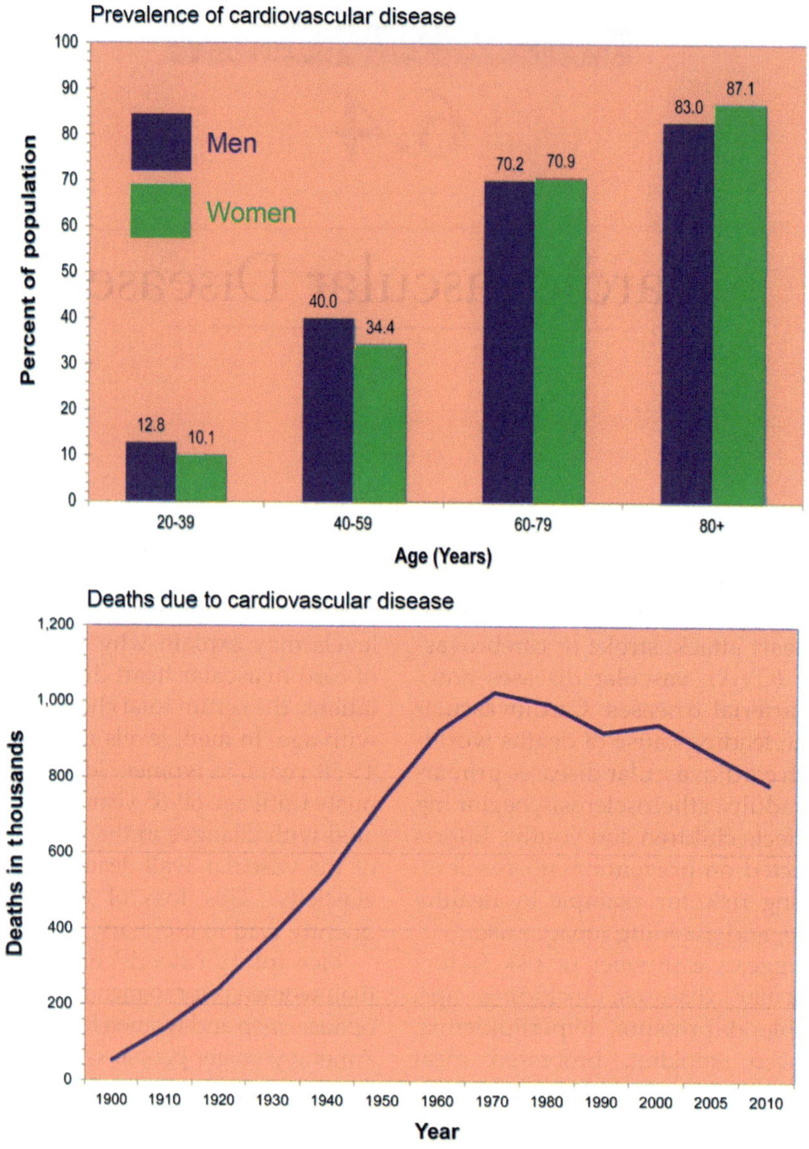

FIGURE 6.4.1 Incidence and death statistics published by the American Heart Association and the American Stroke Association.

artery. Coronary arteries are unusual in that they do not anastomose, or find alternative blood supply routes when a vessel is blocked. As such, a single coronary artery blockage leads to heart muscle death, leading to heart failure.

Cardiomyopathy is a separate disease of the heart muscle. In cardiomyopathy, the heart muscle becomes enlarged, thick, or rigid. In rare cases, the muscle tissue of the heart is replaced with scar tissue. As cardiomyopathy worsens, the heart becomes weaker and less able to pump blood through the body, leading to heart failure.

Hypertensive heart disease is a disease of the heart secondary to high blood pressure. High blood pressure indicates that the pressure inside the blood vessels is too high. As the heart pumps against this pressure, it must work harder. This causes the heart muscle to thicken.

Cerebrovascular diseases are brain malfunctions related to the blood vessels supplying the brain. Like the heart, the brain does not anastomose or find alternative arterial routes to blocked vessels. Blockage of a cerebrovascular vessel causes brain tissue death and stroke or CVA. Hypertension is a common cause of cerebrovascular disease. It damages the blood vessel lining. Sustained hypertension permanently changes the architecture of the blood vessels making them narrow, stiff, deformed, and more vulnerable to fluctuations in blood pressure. A fall in blood pressure during sleep can lead to a marked reduction in blood flow causing ischemic stroke or CVA. A sudden rise in blood pressure due to excitation can cause tearing of the blood vessels. Cerebrovascular disease affects people who are elderly, have a history of diabetes, smoking, or ischemic heart disease.

Peripheral arterial disease results in the obstruction of large arteries not within the heart or brain. It results from atherosclerosis, inflammatory processes leading embolism, or thrombus formation.

Blood thinning medications and cholesterol-blocking statins all reduce risks for cardiovascular diseases. Blood thinning medications include aspirin. Low-dose aspirin, 50-160 mg once each day, prevents heart attack or myocardial infarction and stroke or CVA by decreasing inflammation, or reducing plaque build-up. Inflamed plaque is more likely to cause a heart attack or stroke. Aspirin fights the inflammation by blocking the action of the enzyme cyclooxygenase, or by making the body less able to produce prostaglandins, the chemicals that facilitate the inflammatory response. Prostaglandins also trigger a series of events that cause blood platelets to clump together and form blood clots. As such, when aspirin inhibits prostaglandins, it inhibits both the inflammatory response and the formation of blood clots.

Other blood thinning drugs including blockers of thrombus formation such as warfarin and other agents that prevent heart attack or myocardial infarction and stroke or CVA.

Statins or inhibitors of hydroxy-methyl-glutaryl-CoA reductase, a key enzyme in cholesterol synthesis, are a type of drug used to lower cholesterol production in individuals at risk for cardiovascular diseases. High cholesterol levels have been clearly associated with cardiovascular disease. Statins have been shown to prevent heart attack or myocardial infarction and stroke or CVA in people by lowering their cholesterol levels.

CHAPTER KEYWORDS

Know these keywords and what each word refers to as well as its function. Read over the list. Are all these keywords familiar?

Aspirin
Atherosclerosis
Cardiomyopathy
Cardiovascular
Cerebrovascular accident
Hyperlipidemia
Ischemic heart disease
Myocardial infarction
Peripheral arterial disease
Statin
Warfarin

INTERNET REFERENCES

Use these references to find out more.
en.wikipedia.org/wiki/Cardiovascular_disease
en.wikipedia.org/wiki/Atherosclerosis

www.nhlbi.nih.gov/health/health-topics/topics/atherosclerosis/
www.webmd.com/heart-disease/guide/diseases-cardiovascular
en.wikipedia.org/wiki/Myocardial_infarction
en.wikipedia.org/wiki/Cardiovascular_disease

CHAPTER QUIZ

Answer these questions. Do you know the answers, holy cow! Check yourself out in this quiz, and prove that you are a humanology wiz.

1. What are the common medical terms for heart attack and stroke?
   a. Coronary thrombosis and cerebral thrombosis
   b. Myocardial infarction and cerebrovascular accident
   c. Myocardial infarction and neurovascular accident
   d. Coronary thrombosis and neurovascular accident
   e. Coronary infarction and cerebral thrombosis
2. A drug that reduces cholesterol production is an inhibitor of:
   a. Hydroxy-methyl-glutary-CoA lyase
   b. Inhibitor of hydroxy-methyl-glutary-CoA dehydrogenase
   c. Hydroxy-methyl-glutaryl-CoA reductase
   d. Cholesterol side chain cleavage enzyme
   e. Cholesterol synthase enzyme
3. Aspirin function to prevent heart attacks and strokes by:
   a. Blocking cryptooxygenase preventing prostaglandin synthesis
   b. Blocking cryptooxygenase preventing prostacyclin synthesis
   c. Blocking cyclooxygenase preventing prostaglandin synthesis
   d. Promoting cyclooxygenase preventing prostaglandin synthesis
   e. Promoting cryptooxygenase preventing prostaglandin synthesis
4. Peripheral arterial disease is:
   a. Blockage of a peripheral coronary artery in the heart
   b. Blockage of a major artery in the brain
   c. Obstruction of large arteries in legs and arms
   d. Blockage of a vein in the head
   e. Blockage of a peripheral coronary vein in the heart

CHAPTER

6.5

Hydatidiform Mole

After examining some of the most common types of diseases to affect humans, like bacterial and viral infection, and some of the biggest killers of humans, like cardiovascular diseases and cancers, we examine one unusual remote category of diseases, hydatidiform mole. Hydatidiform mole is only described here because it is a very strange or odd type of disorder, where two sperm or two paternal haploid sets of chromosomes fertilize a single egg (ovum), or an empty egg (an anucleate ovum lacking maternal chromosomes), leading to a living being or a monster.

COMPLETE HYDATIDIFORM MOLE

A complete hydatidiform mole is a pregnancy composed of only placenta tissue, villous placenta tissue (see Chapter 3.10), or trophoblastic tissue. No fetal elements of any type are present. A complete hydatidiform mole takes on a unique anatomy of its own. It presents as hundreds of balls or bloody cysts, hygromatous cysts, as shown in Figure 6.5.1. Each hygromatous cyst or ball arises by distention of the placental tissue by fluids. At hysterectomy, the complete hydatidiform mole may appear like a "bunch of red grapes" or a "bowl of cherries," or a bloody mess as shown in Figure 6.5.1.

The complete hydatidiform mole pregnancy comprises mostly hygromatous cysts (Figure 6.5.1). Normally a woman will look unduly large for her gestation age of pregnancy. The human chorionic gonadotropin (hCG) levels can be 2- to 20-fold higher than those normally seen in pregnancy. This is due to the sheer mass of placental trophoblastic tissue in a complete hydatidiform mole.

Normally, a complete hydatidiform mole may be first discovered by ultrasound examination. The ultrasound shows no hint of a fetus (Figure 6.5.2). It yields a snowflake-like pattern due to the cystic trophoblast tissue upon ultrasound. A complete hydatidiform mole is normally aborted and removed by dilation and curettage, commonly at or around 8-16 weeks of gestation. Cytogenetics of the aborted tissues are required to confirm that the molar object is factually a complete hydatidiform mole.

What is a complete hydatidiform mole? How does biology generate a human being appearing like a bunch of grapes or a bowl of cherries? Multiple genetics experts have examined complete hydatidiform moles. As found, most complete hydatidiform moles are female and have 46,XX genetics. Amazingly, research has shown

http://dx.doi.org/10.1016/B978-0-12-803699-0.00043-8

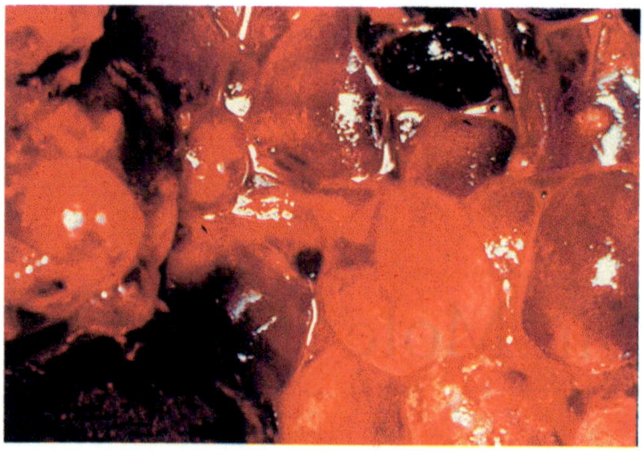

FIGURE 6.5.1 A complete hydatidiform mole as seen upon opening uterus after hysterectomy.

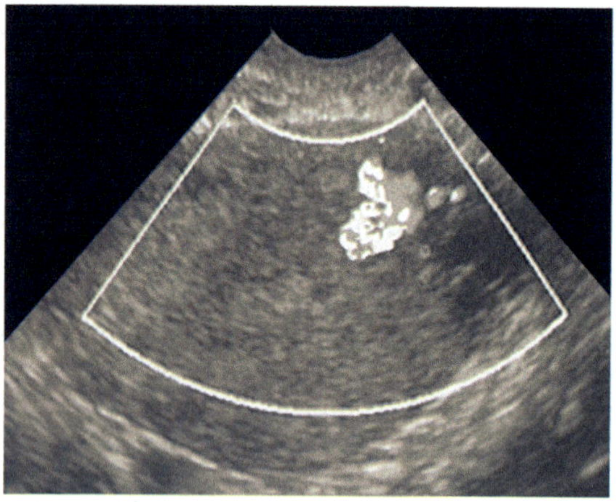

FIGURE 6.5.2 A complete hydatidiform mole at ultrasound.

that complete hydatidiform moles are always diploid. As demonstrated, they comprise two haploid sets of 100% androgenic or paternal origin chromosomes. No chromosomal element of this living being was shown to be of maternal origin or to come from the mother.

Since all diploid chromosomes (2 X haploid) in a complete hydatidiform mole are of paternal origin, it has been inferred that a complete hydatidiform mole must come from an anucleate ovum or an ovum lacking the maternal haploid set of chromosomes. Interestingly, a hydatidiform mole has strictly paternal chromosomal DNA, but the mitochondrial DNA is of maternal origin, suggesting that at least one DNA element is of maternal origin.

It has been speculated that the anucleate or chromosome-free ovum is generated from a meiosis-generated haploid set that has been discharged as a polar body, or from destruction of the haploid chromosome set in either the first stage of meiosis in the ovary or second stage of meiosis in the ovum. Other possibilities include failure of luteinizing hormone (LH) and pituitary sulfated hCG to appropriately stimulate the first stage of meiosis, as they should do as part of the multistep ovarian ovulation procedure.

Figure 6.5.3 illustrates fertilization possibilities considering the generation of an anucleate ovum, which could generate a diploid complete hydatidiform mole. As shown, a single sperm may fertilize an anucleate ovum. The haploid chromosomes are then duplicated. This forms two paternal origin pronucleuses, as would the haploid sets normally from a sperm and an ovum, and then forms a diploid fertilized being (Figure 6.5.3). Alternatively, an ovum may be fertilized by two separate sperm, each forming separate pronucleus, and together forming a diploid being (Figure 6.5.3).

In most pregnancies, the cross-linking of the ovum shell or zona pellucida by enzymes made by the cortical granules will prevent more than one sperm from penetrating the ovum. It is possible that this cross-linking process fails in an anucleate ovum. Two sperm fertilization must, however, occur because it is the only possible explanation for a 46,XY or male complete hydatidiform mole. Imagine, a life-form resulting from the combination and fusing of the haploid sets from two sperms, or from a duplicated haploid set of one sperm. There is no life-form that results from strictly female sex chromosomes, but complete hydatidiform mole is a strange life-form deriving from strictly male chromosomes.

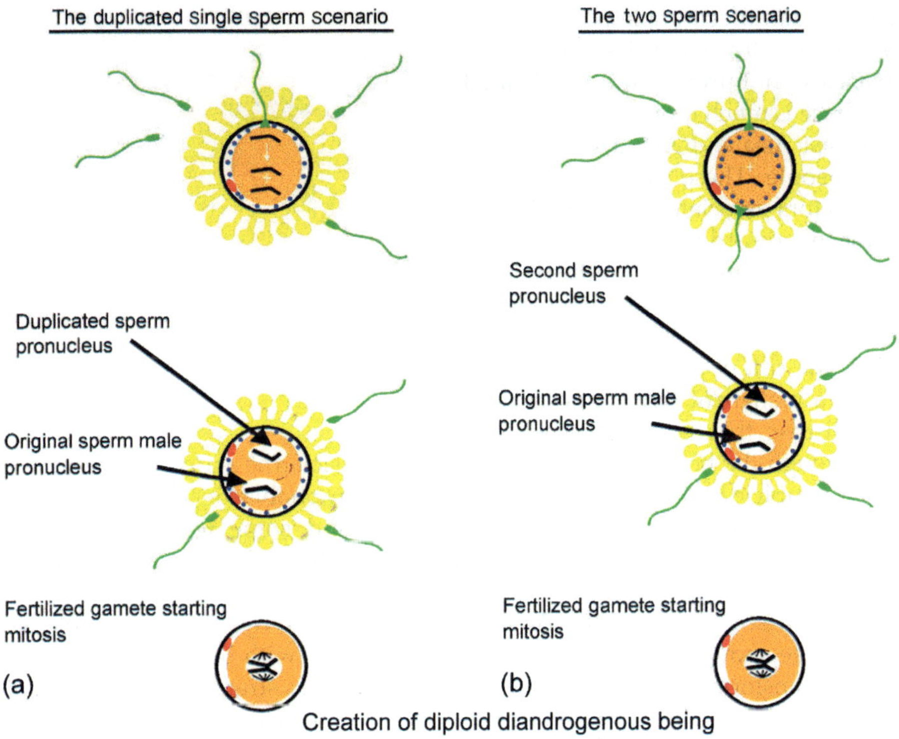

FIGURE 6.5.3 Mono and dispermic pathways for creation of complete hydatidiform mole.

One of the biggest questions is what happens after a totally paternal or diandrogenous fertilized zygote is formed? Cells divide first to 2 cells, 4 cells, 8 cells, 16 cells, and beyond, but how do we end up with a being that is only placenta tissue and no fetal tissue of any kind? The zygote divides, forming 2 identical cells, 4 identical cells, 8 identical cells, and 16 identical cells in the normal way. Then a morula is formed, which normally divides cells into an amnion (fetal tissue) and a chorion (placental tissue). This obviously does not happen appropriately in a complete mole situation. Clearly, in the diandrogenous complete hydatidiform mole case, the morula fails to divide the tissue properly into an amnion (fetus) and chorion (placenta) and by default becomes primarily 100% chorion. It is thought that possibly the maternal set of chromosomes, possibly the maternal X chromosome, signals this division. But clearly there is no signal in the complete mole case. The chorionic being divides further to form a chorion-only blastocyst with no fetal mass, which then implants in the uterus.

It has often been asked whether it is possible that some diandrogenous zygotes make it through the morula division, differentiate appropriately, and end up with normal babies. No one has investigated this possibility. It is clearly by this process and failure of morula division that complete hydatidiform mole ends up as a mass of villous placental tissue. Further differentiation beyond the morula must occur to generate the villous structures needed for hemochorial placentation and the hygromatous cysts or for the complete hydatidiform mole to survive.

A complete hydatidiform mole occurs somewhere between 1 in 900 pregnancies to 1 in every 1500 pregnancies in Europe and the United States. Occurrence is much more common, however, in Southeast Asia and Africa, particularly among tribal people. These people seemingly have genetic factors associated with the higher prevalence for this disorder. The assessed incidence among tribal groups in Southeast Asia and Africa is estimated at 1 in 400 pregnancies.

The biggest complication of complete hydatidiform mole is the risk of persistent or invasive disease. This is a malignant form of gestational trophoblastic disease that invades tissues and metastasizes through the body. It is our understanding that a complete hydatidiform mole that the pregnancy implantation mechanism driven by hyperglycosylated hCG. This is the signal that drives growth and enzymatic invasion in pregnancy implantation (see Chapter 3.10). As now proven by multiple groups, hyperglycosylated hCG binds a transforming growth factor-ß (TGFß) receptor, not an LH-hCG receptor, promoting autocrine cell growth, blocking apoptosis, and promoting invasion through metalloproteinases and collagenases in a cancer-like fashion.

Reports vary widely regarding the proportion of complete hydatidiform mole cases that will develop persistent gestational trophoblastic disease and need chemotherapy. Studies suggest that 16-29% of complete mole cases will need chemotherapy for persistent or invasive disease. We introduce hydatidiform mole as a disease that creates a human monster. This claim is made because the bloody mess that a hydatidiform mole looks like makes it a monster. The high risk of persistent disease or malignancy makes it a monster to battle and survive.

PARTIAL HYDATIDIFORM MOLE

A partial hydatidiform mole is a completely independent ailment to complete hydatidiform mole. Partial hydatidiform mole presents to the physician as a mixture of normal villous placental trophoblastic tissue, hygromatous balls or cysts, and some fetal fragments, without an intact fetus. Figure 6.5.4 illustrates a partial hydatidiform mole as seen after hysterectomy.

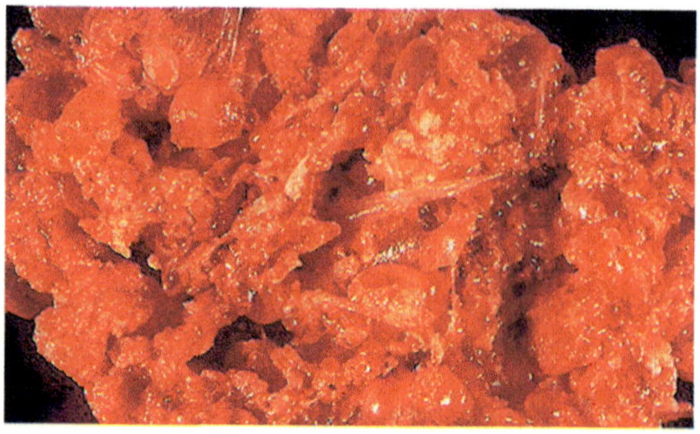

FIGURE 6.5.4 Partial hydatidiform mole at hysterectomy.

A partial hydatidiform mole is a triploid pregnancy with 69 instead of 46 chromosomes and another strange complication of the reproductive process. Fertilization is an intricate process. Part of this process is the cortical granule action (see Chapter 3.10). As illustrated in Figure 6.5.5, upon a single sperm penetration of the zona pellucida or egg shell, the cortical granules enter the perivitelline space and secrete enzymes that cross-link the zona pellucida, making it impenetrable by further sperm. This limits the ovum to penetration by only one sperm.

Partial hydatidiform mole appears to be the result of what happens when this cortical granule process fails and an egg can be penetrated by multiple sperm (Figure 6.5.5). Classically a partial hydatidiform mole is triploid. It could be said that a complete hydatidiform mole is a consequence of an anucleate ovum, and a partial hydatidiform mole is the consequence of failure of the cortical granules to prevent dispermic penetration.

Partial hydatidiform mole can be indicated by ultrasound showing placenta tissue, some hygromatous cysts, and some fetal oddments

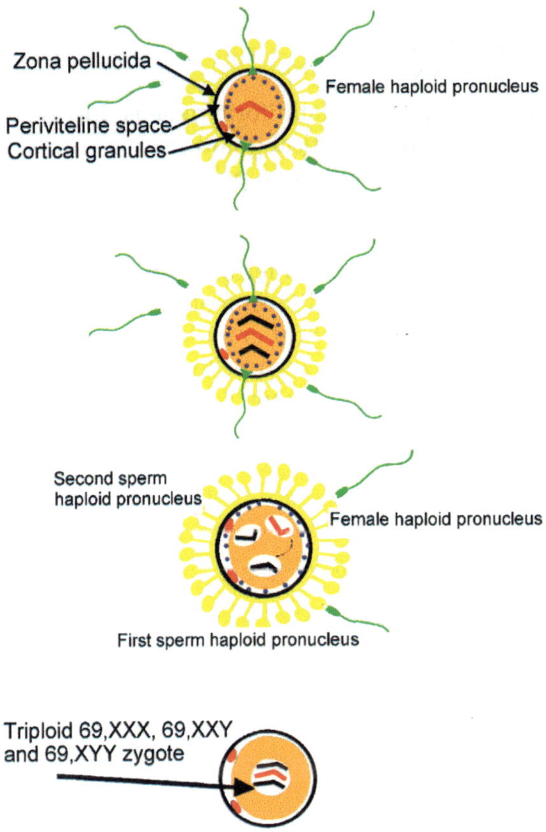

FIGURE 6.5.5 Triploidy of partial hydatidiform mole.

but no clear fetal structure. Ultrasound, however, is strictly an indicator. After spontaneous abortion or dilation and curettage tissue needs to be examined by pathology and cytogenetics determined. If a triploid karyotype is determined, 69,XXX, 69XXY, or 69 XYY, a partial mole can then be confirmed.

Unlike the complete hydatidiform mole, where morula differentiation to a chorion and amnion fails, a partial hydatidiform mole has clear fetal elements. As speculated under complete hydatidiform mole, the presence of the maternal X chromosome in the triploid set may be the signal that limits differentiation. In a partial hydatidiform mole, this maternal X signal is present. However, development of a complete fetus does not happen, suggesting later problems with proper fetal differentiation. We have dissected a few partial hydatidiform mole masses. It can be quite disquieting. One can dissect one's way through cystic and placenta tissues to suddenly come across an intact fully formed finger or an oddball eye.

Partial hydatidiform moles can very occasionally become invasive, developing persistent gestational trophoblastic disease like a complete hydatidiform mole. Research suggests that just 0.5% of partial mole cases become persistent or malignant disease.

Very often, partial hydatidiform moles cause miscarriage of pregnancy or spontaneous abortions. In some countries, dilation and curettage is not carried out when a partial hydatidiform mole is indicated. The partial mole is allowed to remain until the person has a miscarriage.

It is important to always fully differentiate a complete and partial hydatidiform mole properly using cytogenetics. A complete mole is almost always diploid while a partial mole is triploid. A complete hydatidiform mole must be distinguished from a partial hydatidiform mole because of the invasive or persistent nature of the complete hydatidiform mole and not the partial hydatidiform mole.

CHAPTER KEYWORDS

Know these keywords and what each word refers to as well as its function. Read over the list. Are all these keywords familiar?

Complete hydatidiform mole
Cortical granules
Diandrogenous
Gestational trophoblastic disease
Ovum
Partial hydatidiform mole
Persistent hydatidiform mole
Triploid
Zona pellucida

INTERNET REFERENCES

Use these references to find out more.
www.hcglab.com
en.wikipedia.org/wiki/Molar_pregnancy
www.patient.co.uk/health/hydatidiform-mole

CHAPTER QUIZ

Answer these questions. Do you know the answers, holy cow! Check yourself out in this quiz, and prove that you are a humanology wiz.

1. Which karyotype is a partial hydatidiform mole?
   - **a.** 46,XY
   - **b.** 23,XY
   - **c.** 40,XY
   - **d.** 40,XXY
   - **e.** 69,XXY
2. A complete hydatidiform mole is due to:
   - **a.** Failure of cortical granules
   - **b.** Failure of the zona pellucida
   - **c.** An anucleate ovum
   - **d.** Excessive sperm
   - **e.** Failure of sperm fertilization
3. A partial hydatidiform mole is due to:
   - **a.** Failure of cortical granules
   - **b.** Failure of the zona pellucida
   - **c.** An anucleate ovum
   - **d.** Excessive sperm
   - **e.** Failure of sperm fertilization

CHAPTER

6.6

Diabetes

Diabetes is an endocrine disease that affects many Americans today. Approximately 9.3% of the American public has diabetes Type 1 or Type 2, or 25.9% of individuals over 65 years old have diabetes. There are two very clearly defined forms of diabetes mellitus. Type 1 diabetes mellitus results from the body's failure to produce sufficient insulin.

Type 1 diabetes mellitus is typified by loss of the insulin-producing beta cells of the islets of Langerhans in the pancreas, leading to insulin deficiency. Most Type 1 diabetes cases are of an immune-mediated nature, in which T-cell-mediate autoimmune attack leading to the loss of beta cells of the isles of Langerhans and thus insulin. People affected are usually otherwise healthy and of a healthy weight when onset occurs.

Type 1 diabetes is partly inherited, with multiple genes known to influence the risk of diabetes. The onset of diabetes can be triggered by one or more environmental factors, such as a viral infection or diet. There is some evidence that suggests an association between Type 1 diabetes and Coxsackie B4 virus infections. Unlike Type 2 diabetes, the onset of Type 1 diabetes is unrelated to lifestyle. Type 1 diabetes is most commonly treated with daily shots of insulin.

Type 2 diabetes mellitus involves insulin resistance, a condition in which cells fail to respond to insulin properly, which may be combined with relatively reduced insulin secretion. The defective responsiveness of body tissues to insulin is believed to involve the cell insulin receptor. In the early stage of Type 2 diabetes mellitus, the predominant abnormality is reduced insulin sensitivity. At this stage, hyperglycemia or unduly high serum glucose can be reversed by a variety of medications that improve insulin sensitivity or reduce glucose production by the liver.

Diabetes occurs most commonly among American Indians and Alaskan Natives (15.9% of the population) and among non-Hispanic blacks (13.2%) and Hispanics (12.8%). The incidence among non-Hispanic whites is 7.6% (Figure 6.6.1).

Many lifestyle factors are known to contribute to the development of Type 2 diabetes, including obesity, lack of physical activity, poor diet, stress, and urbanization. Excess body fat explains 30% of cases in those of Chinese and Japanese descent, and 60-80% of cases in those of Western and African descent. Dietary factors also influence the risk of developing Type 2 diabetes mellitus. Consumption of

Human Physiology, Biochemistry and Basic Medicine
http://dx.doi.org/10.1016/B978-0-12-803699-0.00044-X

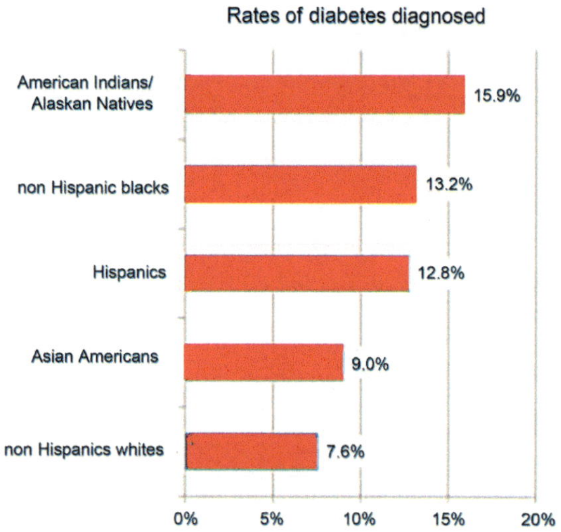

FIGURE 6.6.1 Incidence as reported by American Diabetes Association.

sugar-sweetened drinks is part of the risk. The type of fats in the diet is also important, with saturated fats increasing the risk. Eating excessive white rice appears to also have a role in increasing risk. A lack of exercise is believed to be the cause in 7% of cases.

Insulin is the main hormone that controls uptake of glucose from the blood into the cells of the body. Thus, shortage of insulin or the insensitivity of its receptors causes the disease. The body obtains glucose from four sites, the absorption of food, the breakdown of glycogen, stored glucose found in the liver, and gluconeogenesis or the generation of glucose from other substrates in the body, such as fats and amino acids. Insulin plays a critical role in balancing body glucose levels.

Insulin is made and secreted by beta cells of the islets of Langerhans, in response to rising levels of blood glucose, after eating. Insulin is used by most of the body's cells to absorb glucose from the blood as a fuel, for synthesis of amino acids and fats, or for storage. If the amount of insulin is insufficient, or if cells respond poorly to the effects of insulin, as in insulin resistance, then glucose will not be absorbed properly by the body cells. The effect is persistently high levels of blood glucose and other metabolic disorders.

Diabetes mellitus is a lifetime disease for which there is no known cure. Management of cases must focus on keeping blood sugar levels as normal as possible. This can normally be achieved with diet, exercise, and use of injected insulin and other blood sugar-controlling drugs. Learning about the disease and being involved in the treatment is vital for patients with diabetes. The extreme complications of diabetes are much less common and less severe in patients with well-managed blood sugar dietary levels.

Both Type 1 and Type 2 diabetes have long-term risk complications, which may develop slowly over 10-30 years, but may also be the first symptom in those newly diagnosed with diabetes. The major complications relate to damage to blood vessels. Diabetes doubles the risk of cardiovascular disease. About 75% of diabetic deaths are due to coronary artery disease. Other related diseases include stroke and peripheral vascular disease. Other complications of diabetes include damage to the eyes, kidneys, and nerves.

CHAPTER KEYWORDS

Know these keywords and what each word refers to as well as its function. Read over the list. Are all these keywords familiar?

Diabetes
Hyperglycemia
Insulin
Insulin-resistance
Islets of Langerhans
Type 1
Type 2

INTERNET REFERENCES

Use these references to find out more.
en.wikipedia.org/wiki/Insulin
en.wikipedia.org/wiki/Diabetes
www.diabetes.org/diabetes-basics/type-1/?loc=hottopics
www.diabetes.org/diabetes-basics/type-2/?loc=hottopics
www.medicalnewstoday.com/info/diabetes/

CHAPTER QUIZ

Answer these questions. Do you know the answers, holy cow! Check yourself out in this quiz, and prove that you are a humanology wiz.

1. Why are Type 1 and Type 2 diabetes independent diseases?
   - **a.** Because Type 1 is an endocrine problem and Type 2 is a glucose breakdown problem
   - **b.** Because Type 1 is a problem with the islets of Langerhans and Type 2 is a glucose breakdown problem
   - **c.** Because Type 1 is a problem with the islets of Langerhans and Type 2 is a problem with the pituitary gland
   - **d.** Because Type 1 is a problem at the islets of Langerhans and Type 2 is generally an insulin receptor problem at target tissues
   - **e.** Because Type 1 is a glucose breakdown problem and Type 2 is a glucose storage problem
2. Why does diabetes raise your risk of having a heart attack?
   - **a.** Because diabetes destroys coronary arteries
   - **b.** Because diabetes can cause blockages of coronary arteries
   - **c.** Because diabetes promotes formation of thrombus
   - **d.** Because diabetes damages blood vessels
   - **e.** Because diabetes promotes cholesterol production

CHAPTER

6.7

Alzheimer's Disease

Alzheimer's disease is a permanent, advancing brain disease that slowly destroys memory and thinking skills. Finally, even the ability to carry out the simplest tasks of daily living become near impossible. Alzheimer's commonly first appears after age 60. Alzheimer's disease is the most frequent cause of dementia among older men and women.

Alzheimer's is named after Dr. Alois Alzheimer. In 1906, Dr. Alzheimer found changes in the brain tissue of a woman who had died of an unusual mental illness. Her symptoms were memory loss, language problems, and behavior problems. After her death, he examined her brain and found abnormal clumps, now called amyloid plaques, and tangled bundles of fibers, now called neurofibrillary tangles. These plaques and tangles in the brain are two of the principal features of Alzheimer's disease. The third is the loss of contacts between neurons in the brain. While treatment can help manage symptoms in some people, currently there is no known cure for this devastating disease.

While we don't know how the Alzheimer's disease process begins, it appears likely that damage to the brain commences a decade or more before problems become evident. During the pre-disease stages of Alzheimer's disease, toxic changes are taking place in the brain. Irregular deposits of proteins form amyloid plaques and tau tangles in the brain, and neurons start to work less efficiently. Over time, neurons lose their ability to function and connect with each other, and finally they die.

Inevitably, the injury spreads to a nearby structure in the brain called the hippocampus, essential for memory. After more neurons die, brain areas begin to shrink. In the last stages of Alzheimer's, damage is widespread, and brain tissue has widely shrunk.

Estimations vary, but experts indicate that 5.1 million Americans aged 65 and older have Alzheimer's. Unless Alzheimer's can be properly treated or prevented, the number of people with it will increase significantly. This is because the risk of Alzheimer's inclines with age, and the population of the United States is aging. The magnitude of people with Alzheimer's doubles every 5 year beyond age 65.

Alzheimer's progresses in three stages. There is an early, preclinical stage with no symptoms (Table 6.7.1), a middle stage of mild cognitive impairment, and a final stage of Alzheimer's dementia (Table 6.7.1). The period from diagnosis to death varies from 3-4 years if the person is older than 80 when diagnosed, to as long as 10 years or longer if the person is younger.

Human Physiology, Biochemistry and Basic Medicine
http://dx.doi.org/10.1016/B978-0-12-803699-0.00045-1

TABLE 6.7.1 Stages of Alzheimer's Disease

Early stage Alzheimer's
Absent-minded
Forgetting appointments
Slight changes seen by loved ones
Some confusion
Middle stage Alzheimer's
Deeper difficulty remembering recently learned information
Deepening confusion in many circumstances
Speech impairment
Repeatedly initiating the same conversation
Late stage Alzheimer's
More aggressive or passive
Some loss of self-awareness
Debilitating cognitive deficit
More abusive, anxious, or paranoid

Dementia is the loss of cognitive functioning, the inability to remember, think, and reason, and behavioral abnormalities that interfere with a person's life and activities. Dementia varies in severity from mild dementia, when it is just beginning to affect a person's working life, to the most severe dementia, when a person must depend totally on others for the basics of daily living. Multiple diseases cause dementia, Alzheimer's disease, and vascular dementia, which is caused by a series of strokes or interruptions in the brain's blood supply.

Emotional problems, stress, anxiety, or depression, can make a person more forgetful and may be mistaken for dementia. For instance, someone who has just retired or who is dealing with the death of a spouse may feel sad, lonely, worried, or bored. Dealing with these life changes leaves some men and women confused or forgetful. These problems can be helped by supportive friends and family, but if these feelings last for multiple month, it is necessary to get help from a doctor.

Early-onset Alzheimer's is rare, occurring in men and women aged 30 to 60, representing less than 5% of all people with Alzheimer's. Early-onset Alzheimer's cases are mostly familial Alzheimer's disease, caused by changes in one of three known inherited genes.

Most people with Alzheimer's have late-onset Alzheimer's, which develops after age 60. Studies link the apolipoprotein E gene to late-onset Alzheimer's. This gene has several forms. One of them, apolipoprotein E ε4, increases a person's risk of getting Alzheimer's. Carrying the apolipoprotein E gene does not guarantee, however, that a person will develop Alzheimer's, and some people get this disease without carrying the apolipoprotein E gene. Experts believe that additional genes may influence late-onset Alzheimer's. Scientists around the world have identified a number of common genes in addition to apolipoprotein E that increase a person's risk for late-onset Alzheimer's.

Research also suggests that many factors beyond genetics may play a role in the development of Alzheimer's disease. There is an association between cognitive decline and vascular conditions such as heart disease, stroke, high blood pressure, diabetes, and obesity. Further, a nutritious diet and physical activity can all help people stay healthy as they age. Research suggests that these and other factors might also help to reduce the risk of Alzheimer's.

Alzheimer's can only be absolutely diagnosed after death, by examination of brain tissue and pathology in an autopsy. Doctors now have tools to help them determine quite accurately whether a person has possible Alzheimer's dementia or probable Alzheimer's dementia. The physician may ask about overall health, ability to perform daily activities, and changes in behavior and personality, conduct tests of memory, problem solving, and language, and perform computed tomography or magnetic resonance imaging and try to distinguish using the images of Alzheimer's disease from possible stroke or tumor.

CHAPTER KEYWORDS

Know these keywords and what each word refers to as well as its function. Read over the list. Are all these keywords familiar?

Alzheimer's
Alzheimer's disease
Amyloid plaques
Apolipoprotein E
Dementia
Neurofibrillary tangles
Tangled bundles of fibers
Vascular dementia

INTERNET REFERENCES

Use these references to find out more.
http://www.alz.org/alzheimers_disease_what_is_alzheimers.asp
http://www.nia.nih.gov/alzheimers/publication/alzheimers-disease-fact-sheet
http://en.wikipedia.org/wiki/Alzheimer%27s_disease

CHAPTER QUIZ

Answer these questions. Do you know the answers, holy cow! Check yourself out in this quiz, and prove that you are a humanology wiz.

1. Alzheimer's disease affects approximately how many U.S. citizens?
  a. 3.2 million
  b. 4.3
  c. 5.1 million
  d. 6.6 million
  e. 6.8 million

2. A pathologist examines a dead person's brain; some signs of Alzheimer's might include:
  a. Shrunk tissue, amyloid plaques, tau tangles, damaged hippocampus
  b. Shrunk tissue, damaged hypothalamus, infarcted thalamus
  c. Shrunk tissue, damaged thalamus, lipocomplex deposits
  d. Shrunk tissue, damaged thalamus, apolipoprtein E deposits
  e. Shrunk tissue, damaged thalamus, amyloid plaques

CHAPTER

6.8

Ovarian Cystic Disorders

There are many benign and cancerous forms of ovarian cyst. We examine the biology of the common forms of ovarian cyst, follicular cyst, corpus luteum cyst, and polycystic ovary syndrome.

FOLLICULAR CYST

Follicular cysts are small fluid-filled sacs that develop in a woman's ovaries. A follicular cyst can range from being as small as a pea, to being larger than an orange. Ovarian cysts are generally functional in nature and harmless. In the United States, follicular cysts are found in nearly all women having a normal menstrual cycle. They most commonly occur during a woman's childbearing years.

Follicular cyst is an exaggeration of circumstances occurring before monthly ovulation, promotion of follicular growth (Figure 6.8.1). This type of cyst forms when ovulation does not occur. Commonly, ovulation is blocked by an inadequate supply of luteinizing hormone (LH) at the time of the LH peak. LH normally promotes the sequence of ovulation-related events, follicular growth, meiosis, growth of a stigma, protease permeation of the stigma, and expulsion of the oocyte (see Chapter 3.9). A follicular cyst can result from failure to ovulate due to excess production of follicle stimulating hormone (FSH) at the time of ovulation. If the follicle does not rupture or release its egg, it instead becomes a cyst.

In each month following the ovulation-failed menstrual cycle, FSH, inhibin, and LH may promote the growth of the cyst as they would normally promote the growth follicles (Figure 6.8.1). This is lined by granulosa and theca cells, and is filled with clear fluid. Cysts occasionally become externally palpable and cause concern. Overall, as long as the cyst does not exceed the size of tennis ball, or about 7 cm, it is unlikely to be a malignancy. Fallopian cysts are the most common type of ovarian cyst. They often disappear by themselves and seldom require treatment. Cysts that become very large or last longer than a few months should be removed or tested to determine if they are cancer. Follicular cysts normally disappear naturally over 1 to 3 months. They are benign and have no obvious medical consequence. They may be diagnosed coincidentally during a woman's pelvic examination. All women can have follicular cysts at some point.

Ultrasound is normally used to determine the best treatment strategy. Treatment depends on the size of the cyst and its appearance on

Human Physiology, Biochemistry and Basic Medicine
http://dx.doi.org/10.1016/B978-0-12-803699-0.00046-3

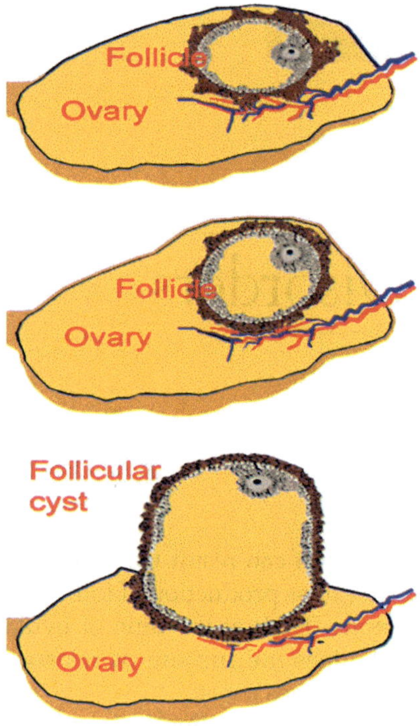

FIGURE 6.8.1 Growth of unovulated follicle into follicular cyst.

ultrasound. Treatment may consist of just observation, or if the follicular cyst is large it can be surgically removed by laparoscopic surgery.

Follicular cysts can interfere with fertility. If a woman does not ovulate and does not produce a corpus luteum, it can block fertility and continue to block fertility as long as there is a follicular cyst each menstrual cycle.

CORPUS LUTEUM CYST

Another type of functional cyst is called a corpus luteum cyst. Corpus luteum cysts occur less frequently than follicular cysts, but can be more complicated. Determining whether a cyst is corpus luteal or follicular is dependent on what is seen by the ultrasound as well as by studying symptoms and signs. Spotting following a missed period, pelvic pain, and a tender ovarian mass indicate a corpus luteum cyst. If a corpus luteum cyst ruptures, the bleeding and pain may necessitate urgent laparoscopic surgery.

Corpus luteal cyst occurs when a corpus luteum is not properly formed from the follicle after the LH peak. If the corpus luteum forms a sealed structure rather than an open structure, and hemorrhages, a cyst is formed. Some corpus luteal cysts happen in cases with sufficient LH to promote ovulation, but insufficient to stimulate proper corpus luteum formation. A corpus luteal cyst produces progesterone just like a normal corpus luteum. This progesterone can feed back and block hypothalamic gonadotropin releasing hormone (GnRH) production, blocking LH and FSH production and in turn causing infertility. Corpus luteal cysts may grow as large as 4 in. in diameter. Cysts normally exist for up to 3 months and normally disappear with time or can be removed by laparoscopic surgery.

POLYCYSTIC OVARY SYNDROME

Polycystic ovary syndrome is characterized by obesity, hirsutism, infertility, and oligomenorrhea. In polycystic ovary syndrome, the ovaries become filled with cysts (Figure 6.8.2).

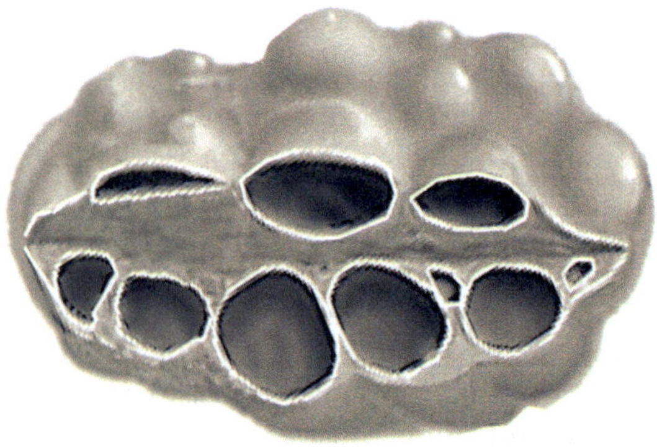

FIGURE 6.8.2 A polycystic ovary.

Under the influence of rising LH and depressed levels of FSH, polycystic ovary syndrome occurs. The low FSH prevents proper maturation of follicles, causing numerous follicles to be developed, all of which become cysts. LH promotes growth of the cysts, thus the polycystic appearance (Figure 6.8.2). LH normally promotes androstenedione production by theca cells; FSH normally aromatizes androstenedione in granulosa cells to form estradiol. In polycystic ovary syndrome with the presence of LH and absence of FSH, excessive androstenedione and androgens (androstenedione is readily converted to testosterone) are produced and minimal estradiol is made. The high blood levels of androgens lead to hirsutism. Without an estrogen peak, no LH peak occurs and ovulation does not occur, leading to infertility. Polycystic ovary syndrome is a common cause of infertility. With ovulation not occurring, a corpus luteum is not produced and no progesterone is produced.

In polycystic ovary syndrome, birth control pills inhibit GnRH production and therefore inhibit LH and stop excessive androgen production. The drug Clomid, clomiphene citrate, promotes ovulation, leading to ovulation of cystic follicles. Laparoscopic drilling surgery makes holes in the cyst and can dramatically help with ovarian androgen production. Normally women are treated with Clomid and birth control pills first. If these do not help, then laparoscopic drilling surgery is considered.

CHAPTER KEYWORDS

Know these keywords and what each word refers to as well as its function. Read over the list. Are all these keywords familiar?

Androgen
Androstendione
Clomid
Corpus luteum
Corpus luteum cyst
Estradiol
Follicle-stimulating hormone
Follicular cyst
Gonadotropin-releasing hormone
Hirsutism
Inhibin
Laparoscope
Laparoscopic surgery
Luteinizing hormone
Ovulation
Polycystic ovary
Progesterone

INTERNET REFERENCES

Use these references to find out more.

www.webmd.com/women/guide/ovarian-cysts
www.mayoclinic.org/diseases-conditions/ovarian-cysts/basics/definition/con-20019937
www.emedicinehealth.com/ovarian_cysts/page2_em.htm#types_of_ovarian_cysts
www.nlm.nih.gov/medlineplus/ency/article/001504.htm

CHAPTER QUIZ

Answer these questions. Do you know the answers, holy cow! Check yourself out in this quiz, and prove that you are a humanology wiz.

1. A follicular cyst is caused by:
   a. Inappropriate LH promotion of corpus luteum formation
   b. An over grown follicle
   c. An unovulated oocyte
   d. An exploding follicle
   e. An improperly developed follicle
2. A corpus luteum cyst is caused by:
   a. Inappropriate LH promotion of corpus luteum formation
   b. An over grown follicle
   c. An unovulated oocyte
   d. An exploding follicle
   e. An improperly developed follicle

Answers to Chapter Quizzes

Section 1, Chapter 1.1 The Cell

Q1. c
Q2. a
Q3. b
Q4. b
Q5. b

Section 1, Chapter 1.2 Human chemistry

Q1. b
Q2. a

Section 1, Chapter 1.3 Sugars, Fatty Acids and Energy Biochemistry

Q1. e
Q2. a
Q3. d
Q4. a
Q5. a

Section 1, Chapter 1.4 Amino Acid Metabolism

Q1. c
Q2. e
Q3. a
Q4. c
Q5. a

Section 1, Chapter 1.5 DNA and Human Genetics

Q1. c
Q2. a
Q3. b
Q4. b
Q5. a

Section 1, Chapter 1.6 Protein structure and folding

Q1. d
Q2. a

Section 2, Chapter 2.1 The Skeleton

Q1. c
Q2. a
Q3. b
Q4. b
Q5. e

Section 2, Chapter 2.2 Skin, Ligaments and Muscle

Q1. e
Q2. d
Q3. a
Q4. a
Q5. b

Section 2, Chapter 2.3 Apoptosis, Growth and Aging

Q1. a
Q2. b
Q3. b
Q4. e

Section 3, Chapter 3.1 The Endocrine system

Q1. c
Q2. a
Q3. e
Q4. e
Q5. c

Section 3, Chapter 3.2 The Heart and Lungs

Q1. a
Q2. d
Q3. c
Q4. a
Q5. d

Section 3, Chapter 3.3 The Circulation

Q1. e
Q2. c
Q3. a
Q4. d
Q5. c

Section 3, Chapter 3.4 Brain and Nervous System

Q1. b
Q2. b
Q3. e
Q4. a
Q5. c

Section 3, Chapter 3.5 Seeing, Hearing, Tasting, Smelling

Q1. a
Q2. d
Q3. e
Q4. d
Q5. a

Section 3, Chapter 3.6 The Liver and the Digestive System

Q1. c
Q2. c
Q3. c
Q4. a
Q5. e

Section 3, Chapter 3.7 The Kidney Filtration System

Q1. a
Q2. b

Section 3, Chapter 3.8 Sex and Reproductive Systems: Male

Q1. b
Q2. b
Q3. a
Q4. e
Q5. e

Section 3, Chapter 3.9 Sex and Reproductive Systems: Female

Q1. c
Q2. b
Q3. e
Q4. a
Q5. b

Section 3, Chapter 3.10 Fertilization, Implantation, and Human pregnancy

Q1. b
Q2. b
Q3. c
Q4. c
Q5. e

Section 4, Chapter 4.1 Human evolution from primates

Q1. e
Q2. e
Q3. a
Q4. c
Q5. a

Section 4, Chapter 4.2 Modern human pseudo-evolution

Q1. d
Q2. c
Q3. a

Section 5, Chapter 5.1 Macronutrients

Q1. e
Q2. b
Q3. d
Q4. d

Section 5, Chapter 5.2 Vitamin and dietary essentials

Q1. c
Q2. d
Q3. b
Q4. f

Section 5, Chapter 5.3 Diet and Body Weight

Q1. c
Q2. e
Q3. b
Q4. d
Q5. e

Section 6, Chapter 6.1 Genetic defects

Q1. b
Q2. e
Q3. c
Q4. a
Q5. a

Section 6, Chapter 6.2 Bacteria, Virus, Fungi, and Infectious Diseases

Q1. e
Q2. c
Q3. a
Q4. a

Section 6, Chapter 6.3 Human cancers and carcinogenesis

Q1. c
Q2. b
Q3. d
Q4. a
Q5. b

Section 6, Chapter 6.4 Cardiovascular disease

Q1. b
Q2. c
Q3. c
Q4. c

Section 6, Chapter 6.5 Hydatidiform mole

Q1. e
Q2. c
Q3. a

Section 6, Chapter 6.6 Diabetes

Q1. d
Q2. d

Section 6, Chapter 6.7 Alzheimer's disease

Q1. c
Q2. a

Section 6, Chapter 6.8 Ovarian cyst disorders

Q1. c
Q2. a

Index

Note: Page numbers followed by *f* indicate figures and *t* indicate tables.

R

S

T

U

V

9780128036990